ISBN 978-0-265-96872-7
PIBN 10917109

This book is a reproduction of an important historical work. Forgotten Books uses
state-of-the-art technology to digitally reconstruct the work, preserving the original format
whilst repairing imperfections present in the aged copy. In rare cases, an imperfection in
the original, such as a blemish or missing page, may be replicated in our edition. We do,
however, repair the vast majority of imperfections successfully; any imperfections that
remain are intentionally left to preserve the state of such historical works.

CANADIAN MACHINERY

AND MANUFACTURING NEWS

A weekly newspaper devoted to the manufacturing interests, covering in a practical manner the mechanical, power, foundry and allied fields. Published by the MacLean Publishing Company, Limited, Toronto, Montreal, Winnipeg and London, Eng.

Vol. XIV	Publication Office: Toronto, December 16, 1915	No. 25

Holden-Morgan Mechanical Plug Wrench

For Screwing The Base Plugs Into Shells

Output 120 per hour

One machine with an operator will do the work of four men. Friction device adjustable, and can be set for any required tension and, when set, the pressure applied will not vary from the desired adjustment.

DIRECT DRIVEN, NO COUNTERSHAFT NEEDED

The plug is screwed in and tightened up entirely by mechanical action, therefore eliminating the variations that result from band work.

CAN BE MADE TO HANDLE ANY SIZE SHELL.
Further Details on Request.

THE HOLDEN-MORGAN COMPANY, LIMITED
579 RICHMOND STREET WEST, TORONTO

FOR TAPPING NUTS

The Tap with the

Con-eccentric Land

One-third of the land from the cutting edge is concentric. The remaining two thirds has eccentric relief.

To show more clearly the free-cutting qualities of these taps, a nut slightly smaller than one and one-quarter inches square was drilled for tapping and held lightly by the corners. The nut was then tapped with a 1¼-7 U.S.S. con-eccentric tap, driven through without reversing. (See cut on right.)

So perfect were the cutting qualities and so easily and freely was the work done that this mere shell of a nut was *not even distorted*, to say nothing of being split as would have happened with a tap differently constructed.

Consider what this means to the nut manufacturer from the standpoint of a *saving in power.* This is but one reason why Pratt & Whitney taps are specified by the largest nut manufacturers in the country.

For a number of years we have made taps with this style of relief. They have won out steadily and consistently on merit alone.

Write for our Booklet "The Tap with the Con-eccentric Land."

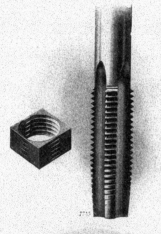

Taps can be sharpened at the only correct point — on the face of the cutting edge.

Sharpening in no way affects their size.

Can be sharpened long after other so-called eccentrically-relieved taps are rendered useless.

We obtain the freest cutting tap with the longest life—an exclusive combination.

Accuracy nowhere else obtainable is secured together with a degree of finish and refinement heretofore thought impossible in a commercial tool.

We want tap users to know why Pratt & Whitney taps are best, also that our claims are based on reasons mechanically correct.

Cut to left is a tapped nut sawed in half showing small area of metal which held it together.

The taps are carried in stock by our Branch Offices and Agencies throughout the world. Prompt service assured to all customers wherever located.

Pratt & Whitney Co. of Canada, Limited

DUNDAS	MONTREAL	WINNIPEG	VANCOUVER
Ontario	723 Drummond Bldg.	1205 McArthur Building	B.C. Equipment Co.

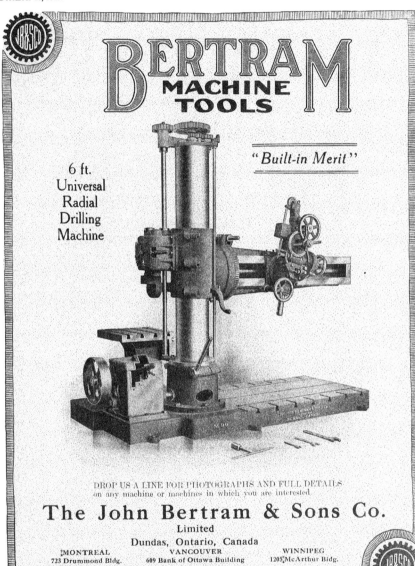
If what you want is not advertised in this issue consult the Buyers' Directory at the back.

The Publisher's Page
By B.G.N.

Your Annual Opportunity
A Message for Both Buyers and Sellers

On December 30th, will be issued a number of CANADIAN MACHINERY which will be of the utmost importance to the buyers and to the sellers of machine shop equipment and supplies.

ANNUAL REVIEW (and 4th Shell) NUMBER

During the past year CANADIAN MACHINERY has been unusually interesting from an editorial standpoint and a series of successes and surprises will culminate in our coming Number. As the majority of our metal workers are engaged directly or indirectly in the manufacture of munitions, it is only natural to expect that our Annual Review (and fourth Shell) Number will deal with Canada's latest industry. Articles of the greatest value to manufacturers and those engaged in shell manufacture will be included in this unusually interesting issue.

This feature will be one of many and the whole issue will teem with valuable information.

Orders for additional copies of this Number will have to be received by December 21st.

TO MANUFACTURERS AND DEALERS

A sales message in our Annual Review Number will be as safe as any gilt-edged investment, and **will pay larger dividends.**

It is significant that the majority of advertisers in our last Annual Review Number have reserved increased space in our coming issue. That can have but one meaning!

No other paper in Canada is more closely read than CANADIAN MACHINERY. To be in our Annual Review Number is to court fortune with a sunny smile. To overlook your opportunity is to deliberately invite neglect during a most important period of industrial activity and development.

Write, wire or telephone your space reservation at once. Send copy as early as possible, and not later than December 21st.

Wise buyers will preserve our Annual Review Number for reference purposes. Wise manufacturers will see that their lines are represented.

CANADIAN MACHINERY
143-153 University Avenue   TORONTO

If what you want is not advertised in this issue consult the Buyers' Directory at the back.

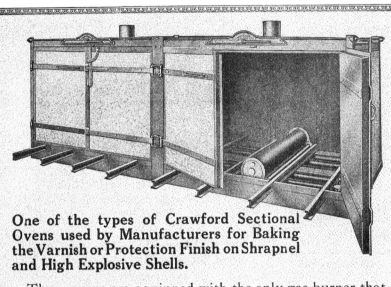

If what you want is not advertised in this issue consult the Buyers' Directory at the back.

"Murchey" Tools

are threading successfully
all types and sizes of

High Explosive Shells

English, French, Italian,
United States and Russian.

LARGE SHELLS

of 9.2' and 12' diameter

are calling for improved and larger types of Tools to produce them.

Murchey Service

which means Murchey Collapsing Taps and Self-opening Dies — is doing this work NOW in a number of the largest munition plants with entirely satisfactory results.

Send us B-P of your requirements and let us quote you on the necessary tools.

Murchey Machine & Tool Company

75 Porter Street

DETROIT, - MICH.

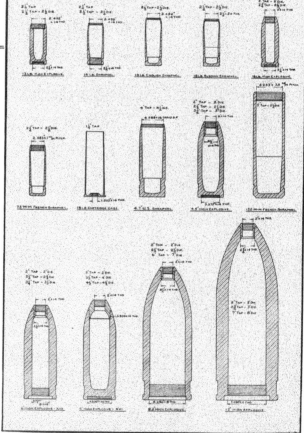

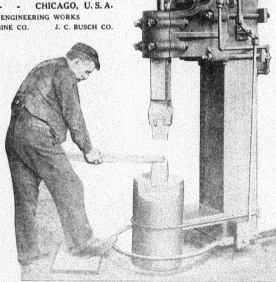

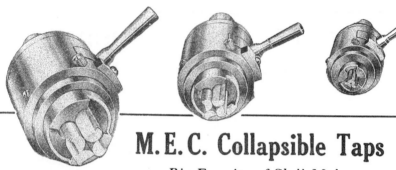
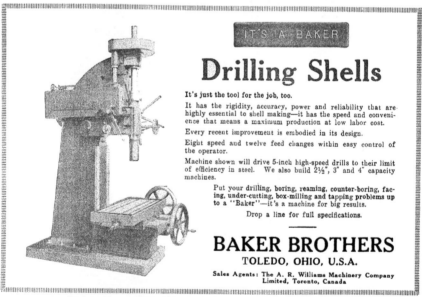

The advertiser would like to know where you saw his advertisement—tell him.

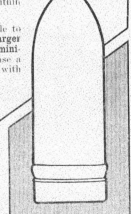
If what you want is not advertised in this issue consult the Buyers' Directory at the back.

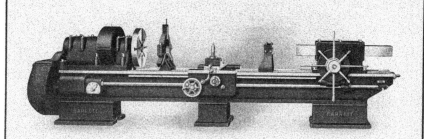
The advertiser would like to know where you saw his advertisement—tell him.

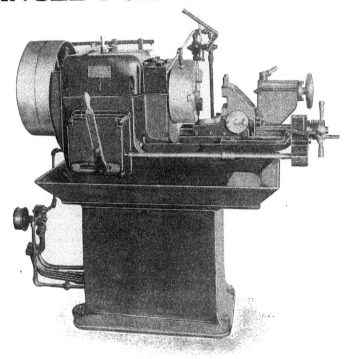
If what you want is not advertised in this issue consult the Buyers' Directory at the back.

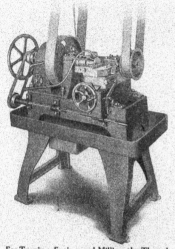

If what you want is not advertised in this issue consult the Buyers' Directory at the back.

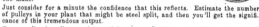

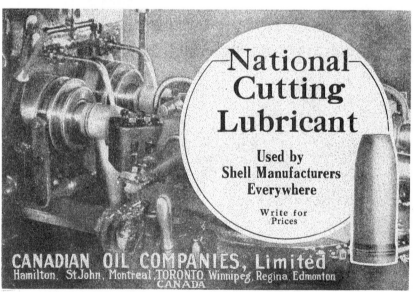
The advertiser would like to know where you saw his advertisement—tell him.

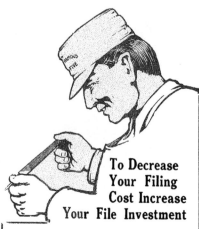

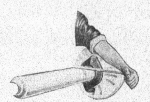
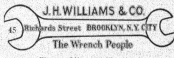

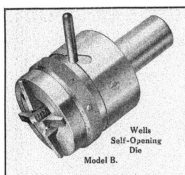

**Wells
Self-Opening
Die
Model B.**

We want to send you the booklet describing the different models. Are you willing to try the W.S.O.D. in your shop under your own conditions?

W. S. O. D.

We call it the "universal die" because there is not a screw-cutting machine manufactured on which it will not fit.

Its very appearance attracts and holds you—you instinctively know it will do the work—and it will.

It is the simplest and most efficient of all automatic opening die heads.

WELLS BROTHERS COMPANY OF CANADA, Limited
GALT   -   ONTARIO

Sales Agents:
The Canadian Fairbanks-Morse Company, Limited,   Montreal, Toronto, Vancouver, Winnipeg, St. John, Calgary.

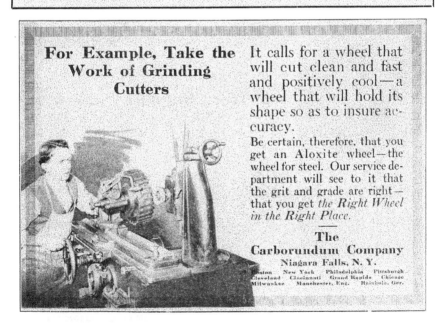

If what you want is not advertised in this issue consult the Buyers' Directory at the back.

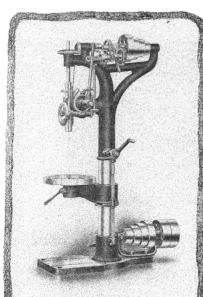

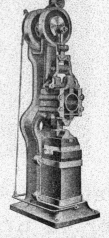

A General Purpose 20" Drill Press

This is a very valuable machine for manufacturing, as it is adapted to a large variety of work; and with the Power Feed and Automatic Stop, with proper jigs, they are run in gangs of from 2 to 6 or more, greatly facilitating and reducing the cost of getting out work.

This Drill has features in the manner of changing from lever to wheel feed that are very quick and effective. By a slight turn of the lever a clutch is thrown from lever to wheel feed, and each is independent of the other when in use.

Full specifications on request

The Canadian Fairbanks - Morse Company, Limited

St. John   Montreal
Quebec   Ottawa
Toronto   Hamilton
Winnipeg   Saskatoon
Calgary   Edmonton
Vancouver   Victoria

Forge Your High-Speed Steel on a Fairbanks Power Hammer

An ideal hammer for this class of work. By the use of a steel spiral spring, the force and weight of the blow is greatly increased without any danger of breakage. This spring also acts as a cushion, relieving the machine of all jar. The head strikes a quick, sharp blow and instantly gets away from the work without chilling the stock.

Made in many sizes with rams from 25 lbs. to 250 lbs.

Write now for full information.

Prices gladly quoted on receipt of specifications.

The Canadian Fairbanks - Morse Company, Limited

St. John   Montreal
Quebec   Ottawa
Toronto   Hamilton
Winnipeg   Saskatoon
Calgary   Edmonton
Vancouver   Victoria

The advertiser would like to know where you saw his advertisement—tell him.

Large Shells : Production Problems and Possibilities---V.

By C. T. D.

In preparing to undertake the production of large shells up to 9.2 in. dia., manufac-
turers will encounter problems of a nature altogether different from those connected with 18
pdr. shells. Automatic machinery will not be so applicable to the larger sizes, and produc-
tive ability will centre largely on such points as sequence of operations, tooling methods, etc.

ALTHOUGH the majority of manu-facturers who have undertaken the production of large shells are already operating plants of more or less magnitude, they are more than likely to find that a considerable number of new machines will be necessary in order to maintain deliveries. Those firms who are already established, especially in lines of a general nature, will lean to-wards the adoption of machines which possess features and quality of more than temporary interest. The increased industrial activity which is ultimately expected justifies the selection of ma-chine tools which can be expected to continue rendering good service for years to come.

Machines of this class have been pro-duced by several makers during past years, which, although primarily adapted to munitions manufacture, are of con-siderable value as producers in estab-lishments of the class referred to.

The installation of such machines at the present moment, while initially more costly, is ultimately a double economy, because, firstly, they are the result of many years' experience by large organi-zations, who have made a success of work of this kind. They have been tested out under months and years of severe service, consequently their makers can supply them with the assurance of immediate production without that ex-perimenting which has so frequently to be done when one is operating along new lines. Secondly, when their work

FIG. 15. PROJECTILE BORING LATHE FOR SHELLS FROM 6 IN. TO 12 IN. DIA.

on war material is finished, their sound design and first-class construction give them a high permanent market value, whether they are retained in a factory or offered for sale.

For many years the John Bertram & Sons Co., of Dundas, Ont., have spe-cialized in the building of machine tools of the highest grade for many lines of engineering work, and their activity in providing equipment for munitions plants has been characterized by all the features of their normal efforts.

The accompanying schedule of opera-tions suggested by them is based on use of machines which have been developed specially for large projectile work, yet have that value for other lines which

appeals to the fore-armed manufacturer. Minor operations are performed on ex-isting types of machines which call for no special comment.

Comparing these operations with pre-vious suggestions, the principal feature observed is the thoroughness of the me-thods adopted. For instance, when fin-ish turning the body in the fifth opera-tion, the shell is mounted on a special arbor which extends through the nose, thus insuring absolute concentricity be-tween inner and outer surfaces. While the careful performance of previously published methods on miscellaneous ma-chines would produce work of the re-quired accuracy, the absolute certainty of these methods ultimately results in greater output.

Illustrations of the principal ma-chines referred to are shown in Figs. 14, 15 and 16. Fig. 14 shows the 36 in. heavy duty engine lathe used for the third operation. While of seemingly conservative design, this machine has several features which call for more than passing mention. The strain of constant heavy cutting on rough forg-ings demands more than ordinary stiff-ness, and both the main spindle and the tailstock spindle are supported in such a manner as to insure the desired rigidity. The drive is from a variable speed motor, with a minimum amount of gearing, which is entirely enclosed. The headstock is of box form, the side walls imparting considerable added stiffness to the spindle bearings.

A very liberal amount of bearing sur-face is provided for the carriage, while the attention given to such important

FIG. 14. 36-IN. MOTOR-DRIVEN LATHE FOR TURNING LARGE SHELLS.

details as lubrication and the provision of dust scrapers to prevent the entrance of cuttings beneath the carriage are

mediate interest at the present moment is the heavy projectile boring lathe, shown in Fig. 15. This machine swings

a most substantial nature. The bed sits flat on the foundation over its entire length, flat tracks and large cross section being prominent features in the design.

The headstock is of box construction, with all gears enclosed, the direct connected motor being mounted on a pedestal above the rear bearing. The front spindle bearing is 8 in. dia. x 23 in. long, these dimensions conveying some idea of the liberal proportions throughout. The bearings and gears, which are entirely enclosed, are lubricated continuously from a high level tank contained in the head, the oil being constantly pumped up from the bottom of the headstock.

The 35 horse-power motor is of the variable speed type, and the gearing through which it drives gives three mechanical changes controlled by handwheels. The electrical control of the motor gives 20 variations, which, combined with the gearing, gives 60 different face plate speeds from 2½ to 77 revs. per min. The control shaft extends along the front of the bed, and the con-

FIG. 16. 30-IN. ENGINE LATHE FOR GROOVING AND BAND TURNING.

points which appeal to the judicious purchaser.

The machine which is of the most im-

36 in., and is suitable for rough and finish boring projectiles from 6 in. to 14 in. diameter. The design throughout is of

Sequence of Machining Operations for 9.2 in. Howitzer Shell Forgings.

Drill Center Hole and Face Nose of Shell

Performed by heavy radial drill. The shell is dropped over an arbor which fits the shell, and which is located central with drill spindle. The twist drill used is a little smaller than the size of the bottom of the thread in the nose, enough stock being left to make a reaming operation. After the hole is drilled twist drill is taken out, and a pilot facing cutter is used, which sweeps off the nose end of the shell, giving a flat surface, a little larger than is required for the finishing end.

Cutting-Off the Open End

For this operation heavy cutting-off machines are used, the shell being held in the spindle of the cutting-off machine, back against a stop, or in some cases a heavy engine lathe is used, to the face-plate of which is attached a heavy split cast iron chuck, open end of which runs in steady rest.

Rough Turning Forgings

This operation is performed on 36" heavy duty engine lathe, and the open end of the shell is placed on an expanding arbor which grips the inside bore, the tailstock having a revolving center, fitting in the hole already drilled in operation 1. This machine is arranged with a profiling attachment, so that the roughing cut will turn the outside of the shell to shape.

Boring Out the Shell

This operation is performed on a 36" projectile boring lathe, which is motor driven. To the faceplate of the lathe is bolted a heavy cast iron split chuck, which is supported on the outer end, by heavy steady rest. The shell is slipped into the chuck, and clamped in position. The projectile boring lathe has a heavy square steel bar, running through the tailstock, and is driven by all steel gearing, and has sufficient power to carry a three fluted roughing reamer, each cutter of which has serrated edges for breaking the chips. After this cutter is driven in the desired depth, it is withdrawn and the second cutter bar takes its place. The second cutter bar is of acorn shape, having two high speed steel cutters of the desired shape to finish the internal bore of the shell. This operation demands a heavy machine, as while the operation is performed in a comparatively short time, the character of the work must be perfect.

Finish Body Turn

This operation is performed on a 30" heavy duty engine lathe, the shell being mounted on a special arbor, conforming to the inside diameter of the shell, with the arbor extending through the shell, and held in position by nuts on either end of the arbor, which is driven from the face-plate. The lathe is provided with a special forming attachment, which enables the operator to not only turn the outside diameter of the shell to an exact size, but also gives a perfect form to the nose of the shell.

Boring and Threading the Nose

For this operation 26" heavy duty lathe is used, the large end of the shell being held in chuck bolted to faceplate of the lathe, and body of the shell running in steady rest. The lathe is provided with a turret on the carriage, enabling the use of a number of tools in rotation for this operation, the hole being reamed out to the proper size for threading, the nose of shell being faced to a finish, the thread being cut by collapsible tap, and a tool being used of suitable shape to face up the inside of the shell back of the thread.

Waving and Undercutting

A 30" heavy duty lathe is used on this operation, or special lathe of heavy design, equipped with a waving and undercutting attachment. This attachment consists of a fixture bolted to the bed of the lathe, and having at the back two tools operated by cams which remove the stock and undercut the corners. The front tool on the fixture has the required number of depressions for the number of waves required, and is operated by a cam on the faceplate which completes the form, giving the waving motion required.

Counterboring and Threading the Base

A heavy duty 30" lathe is used having a large cast iron split chuck secured to the faceplate, with the open end running in steady rest. On the cross-slide of the lathe is located a heavy turret with the necessary tools required for the counterboring operations and facing of the end of shell. The threading is performed by a chaser of the desired pitch, which requires a number of cuts to cut the thread to the desired size.

Pressing the Copper Band to Place

The band being heated is slipped over the shell, and forced into the groove prepared for it, either in a steam hammer, by means of two half circle dies, a few blows only being required to force it into its place, or by an hydraulic press, either method giving very good results.

Turning the Copper Band

This operation is performed on a 30" heavy duty lathe, with special attachment, or a single purpose lathe with special attachment, the tool of the attachment roughing the copper band to an approximate shape, and the other tool finishing the band to an exact size and shape. To the fixture carrying the finishing tool is a special attachment for undercutting the copper band to suit the specifications.

Facing the Base Plug in Position

The base plug being screwed in place, the shell is then chucked in heavy duty 30" lathe, having a split cast iron chuck bolted to the faceplate, open end of which runs in a steady rest. The base plug having been left a little thicker than the depth of counterbore in the shell, the surplus stock is then removed, giving a perfectly flat surface on the base of the shell.

Turning the Base Plug

The base plug is a drop forging of suitable size with ample stock for finishing operations. The forging is chucked, faced and turned to the proper diameter, then threaded to an exact size, leaving the large diameter a little longer than is required to fill the recess in the shell, this surplus stock being taken off as stated in operation 11.

After the above operations have been completed, having been inspected at intervals during the process of manufacture, it is then necessary to varnish the inside and bake same. This leaves a perfect glasslike surface for the high explosive. Before shipment the shells are painted with vaseline, to prevent them from rusting. The work throughout has to be of a very high character, in order to suit the requirements, and conform to the specifications insisted upon.

trol handle is carried by a bracket, which can be moved along the bed as desired, giving complete control over the motor from the position most convenient for the operator.

As boring operations form almost the entire work of this lathe, special provision in the shape of a self-oiling roller type thrust bearing has been provided to prevent trouble under continuous heavy duty.

The tailstock is designed to carry a square boring bar of generous proportions, and made from a steel forging. The bar is fed forward by a pinion engaging with a rack on its underside. The feed shaft extends along the back of the bed. Mounted on the back end of the rack pinion shaft is a large worm wheel, the worm which drives it being driven from the feed shaft through a small train of spur gears.

The movement of the bar is controlled by suitable cutters, which are fed straight into the forging, the travel of the boring bar being in a straight line only.

The engine lathe shown in Fig. 6 is a 30 in. double-geared type, suitable for performing the various operations involved in grooving, waving and bandturning or forming. Although not of such massive design as the lathe shown in Fig. 14, it possesses all the elements of rigidity, the tailstock being liberally proportioned and securely bolted to its base block.

The four-sided tool post, shown in the illustration, is that which is ordinarily supplied. When equipped for grooving and waving, a special rest is substituted, consisting of a lower slide, which is mounted directly in the carriage, and has cross adjustment by a hand screw. A tool slide is mounted on this lower slide.

baud lathe, the drawing showing the arrangement for the 9.2 in. shell.

The bearings are cast integral with the bed, and are of large size, the front one being 9¾ in. dia. x 7 in. long, and the rear 6 in. dia. x 7 in. long. Both bearing caps are machine fitted, and secured by four heavy studs, the forward cap being interlocked with the bearing to take the end thrust without end shift. Bearings are babbitt-lined and provided with shims for taking up wear.

The tailstock is a very rugged casting, located and bolted to the bed to suit the size of shell being machined. The spindle is 5 in. diameter, finished by grinding, and holds a heavy centre fitted to a No. 5 Morse taper. The spindle has a quick movement of about 6 in., actuated by rack and pinion. Its exact position is determined by a swinging stop at the rear, which drops over the end of the spindle when it has advanced far enough.

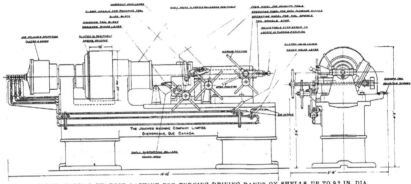

FIG. 17. SPECIAL PURPOSE MACHINE FOR TURNING DRIVING BANDS ON SHELLS UP TO 9.2 IN. DIA.

by the large hand spiders on the tailstock. Eight variations in feed are provided from .005 in. to .030 in. per spindle revolution. The total length of feed of the boring bar is 48 in. at one setting of the tailstock.

In addition to being securely clamped to the bed, the tailstock is fitted with a tail brace which engages with a rack in the centre of the bed, to prevent slipping under excessive thrust pressures.

The forward end of the bar is supported by an A-shaped bracket, which can be adjusted close to the mouth of the forging. This bracket aids greatly in eliminating vibration, and assists in the production of accurate work from irregular forgings. The steady rest is of stiff box section, with four rests, and has a capacity of from 6 in. to 20 in. diameter.

When in use, this machine is fitted with boring heads of any desired style, the curved nose interior being formed

To obtain the waving motion, this tool slide is moved back and forth parallel with the bed by means of an eccentric mounted on a cross shaft, which extends to the back of the cross slide. Bevel gearing drives this cross shaft from a shaft along the back of the lathe. This latter shaft is geared to the main spindle of the head, and suitable change gears are provided for obtaining the number of waves required by the various shells.

While a machine of this type can be utilized for forming the bands, it is not always desirable to use a machine of such capacity on single operation work, except on the largest size shells. For use on shells from 8 in. to 12 in. dia. special band turning machines have been designed, one of which is shown in drawing, Fig. 17.

This machine has been placed on the market by the Jenckes Machine Company, of Sheerbrooke, Que., and is described as a single purpose air actuated

A limited adjustment is provided by screw and lock nut in the stop. Thus the centre acts as a stop and a bearing for the shell, enabling the band to be finished the correct distance from the end.

The driving spindle is steel and hollow, finished inside and outside; the outside by grinding on dead centres. The chuck is built into the extension of the spindle, and is of the collet type, the jaws being opened and closed positively by compressed air. It is operated by an air cylinder at the rear, the controlling valve of which is at the operator's right hand, below the tailstock. In closing the jaws, the shell is pushed forward until stopped by the tail centre. Jaw movement continues until the full pressure of the cylinder is exerted. The drive is through a single clutch pulley, 22 in. diameter, 10½ in. face. The pulley is bronze bushed and runs loose on the spindle. The clutch is the full dia-

(Continued on page 549.)

The Development of Quick Acting Forging Presses--I.

By A. J. Capron, M.Inst. C.E.

The accompanying article formed the subject matter of a paper presented at the International Engineering Congress, San Francisco, last September. Recent progress and present status of the art of forging, together with the relative advantages of the quick-acting forging press over the steam hammer, constitute the salient features of the detail contents.

DURING the past ten years great developments have taken place in the art of forging, and over widening fields of industries, requirements in forgings have increased rapidly. In the first place, large forgings are required, particularly for the increased calibre of guns that have been coming into use, and for turbine drums, wheels, and spindles, in many cases requiring an ingot of as much as 100 tons weight.

Again, forgings are being much more extensively used in preference to steel castings, in marine work, where reliability is the first consideration, and some saving in weight is of more consequence than economy in production. It also applies largely to forgings of all kinds for general engineering work. When steel castings were introduced, they were used in many cases to replace forgings, at first often made in iron and subsequently in steel, the steel castings effecting a great saving in the cost of machining, particularly in the case of articles of intricate form. The introduction of high-speed steel during the last fifteen years has so reduced the cost of machining that forgings have been replacing castings, in many cases advantageously as regards the cost of production, and their use in this direction is continually extending. This development is greatly assisted by the improvements that have been made in the art of forging.

For the production of the heavier forgings, presses have been used in preference to hammers for as long as thirty years in many of the large steel works; but such presses are comparatively slow in their action, and it is only within the last ten years that great improvements have been made in their

design and construction, and more particularly in their speed of working, which has rendered presses suitable for almost all classes of work, except the lightest forgings, which come within the range of small steam hammers or drop-stamps.

Forging Press and Steam Hammer Comparison

The press possesses many advantages over the hammer in its effect on the material, in the manipulation of the forging, in output and economy in working, and in absence of noise and vibration, which, in the case of heavy hammers, is always objectionable, and often detrimental to adjacent buildings and machinery.

Effect on the Material

Dealing first with the effect on the material, the squeeze of the press penetrates to the centre of the forging, as is evidenced by the bulging of the sides of the forging at each stroke; whereas the blow of the hammer has much more of a surface effect, often leaving the sides of the

forging quite concave unless the power of the hammer is very ample. As a result of this difference of action, the hammered forging shows a slightly finer texture on the surface, but a distinctly more open grain of metal towards the centre. The pressed forging, on the other hand, shows a fairly fine and practically uniform texture throughout the entire section.

Individual tests may not always show a very marked difference between pressed and hammered forgings, because a good deal depends on the relative powers of press and hammer to the size of the forging, the temperature at which the work is done, and the subsequent annealing or heat treatment: but in the course of the regular manufacture of tires, for example, it has been found that much better and more reliable tests have been obtained from the press than from the hammer. This advantage of the press is recognized by Government requirements for forgings which stipulate that, for rolled or hammered steel, the ingot must have an initial section eight times the finished section, whereas those pressed need have an initial section of four times only.

Manipulation of the Forging

For the expeditious and accurate manufacture of forgings, a great deal depends on the ease and convenience with which the forging itself and the tools required for its production can be handled; and comparing the action of a press with the blow of a hammer, it will be readily understood that the advantage in these respects is entirely on the side of the press. In the first place, it is essential in working under a hammer that the forging should lie true on the anvil, otherwise a fair blow can-

FIG. 1. MODERN FORGING PRESS OF 6,000 TONS CAPACITY.

not be struck, and an objectionable shock may be caused to the crane carrying the forging, and perhaps also to the men who are manipulating it; whereas a press will bring the forging to its correct position on the anvil without causing any jar and without interfering with the effect of the forging stroke.

When using tools under a hammer, much greater care must be used to place them correctly and to hold them in position, because the repeated blows of the hammer tend to displace them. A press on the other hand, once the tool is correctly placed, holds it in position until the cutting or forging stroke is completed. This enables such work as cutting or necking to be done much more easily and accurately under a press than under a hammer. For making a tapered forging with a press, a wedge-shaped block to give the correct taper can be laid on the anvil without any fixing. With a hammer, such an arrangement would require fixing to prevent its being displaced.

To forge accurately to size with large presses, a scale or indicator-dial can be used, worked from the cross-head, which enables the operator to read the reduction made at each stroke and to work the forging down to the finished size without any measuring. With small presses, a gauge piece is often used between the tool-faces which enables the forging to be made exact to size. With a hammer neither of these methods can be used conveniently.

Output

In almost all classes of work the press has a great advantage over the hammer in the matter of output. The blow of the hammer produces a very limited effect, generally reducing the forging by only the fraction of an inch, whereas the press will make a reduction of several inches per stroke, which, in straightforward cogging or rough forging work, naturally increases the output enormously. A good modern forging-press of 2000 tons will make as much as twenty 3-in. (75-mm.) strokes, or a total penetration or reduction of the forging of 60 in. (1,-500 mm.) per minute. In rounding up or finishing a forging, such a press will work up to 60 strokes a minute, so that it is as fast as a hammer in finishing, and much superior to a hammer in cogging and rough forging.

As an instance of the output obtainable with a modern quick-acting press, a 2000-ton press of the steam intensifier type, starting with a 45-in. (1.143 m.) ingot, has cogged down and finished a 30-ft. (9-m.) length of well-finished 15-in. (381-mm.) shaft in one heat, using only flat tools. With a hammer of equivalent power, this work would have required not less than three heats. As another instance, a 1000-ton Press, working on 24- in. (600-mm.) ingots, has

forged 31 tons of 15-in. (375-mm.) round shafts in eight hours, and as much as 34 tons of miscellaneous forgings have been produced in eight hours by the same press. These are undoubtedly performances which could not be approached by any hammer.

With forgings of a more intricate shape, the advantages of a quick-acting press over a hammer are even more marked, because the same advantage in speed of penetration is obtained combined with much greater facility in handling the forging and, at the same time, the production of a forging more accurate to shape and size. As an instance, an 800-ton press, which is nominally equivalent to a 7-ton hammer, has produced the forgings for an anchor weighing 11 tons in three days. Properly, this size of work should have been done under a press of 1,200 tons, which could have cut the time occupied to about one-half.

Hollow forging is another class of work for which the press is much better adapted than the hammer. This is particularly the case in expanding operations for making hoops or drums, because the mandrel and its supports stand up far better against the squeeze of the press than against the blow of the hammer, resulting in more expeditious and better work.

Economy in Working

The principal factor in economy in working is the much larger amount of work that can be done in a heat under a good quick-acting press than under a hammer, as there is not only the actual saving in time, but considerably fewer heats are required to produce the same weight of forgings, and consequently the expenditure in fuel in reheating is greatly reduced. Besides this, the steam consumption of a good modern press is barely half that of a hammer for the same output.

Actual results over extended periods with presses have shown that in the production of solid, heavy forgings, such as shafts, the consumption of coal for steam production averages 3 cwt. (150 kg.) per ton of forgings; but for general work the consumption may be taken at from 3 cwt. to 5 cwt. (150 kg. to 250 kg.) per ton. It is evident that with a hammer a great deal of useful work is lost in vibrations; and in a paper on "Power Forging," with special reference to forging-presses, by Gerdan and, Mesta, published in the Journal of the American Engineer, July, 1911, it is assumed that this loss amounts to one-third of the useful work, which is probably not over the mark. A calculation by the same authors gives the following results for equivalent powers of hammer and press:

　Steam hammer, per stroke, 8 lb. (3.73 kg.) steam.

Steam, hydraulic press, per stroke, 3.7 lb. (1.68 kg.) steam.

An exact comparison is difficult to obtain; but the statement made above— that the steam consumption of a good modern press is barely half that of a hammer under average working conditions—is probably well within the mark.

When coal-fired furnaces are used for heating the forgings, it is often possible by means of the waste heat from the furnaces to raise sufficient steam for working the press. This is generally more convenient and economical than using gas-fired heating furnaces and independently-fired boilers.

Absence of Noise and Vibration

The press, with its silent working and absence of vibration, contrasts very favorably with the hammer, which causes much inconvenience and often considerable detriment to adjacent furnaces and buildings and to the working of any machine-tools in its vicinity. In many cases the inconvenience caused by hammers to adjacent establishments has been so great that, on this account, it has been necessary to replace them by presses. Besides this, the foundation required for the press is very small and inexpensive compared with what is required for a hammer; and when the nature of the soil is unfavorable, this difference is accentuated. The working of the press is also much less detrimental to the tools which can often be of cheaper construction and require less frequent renewals.

Capacity of Press and Equivalent Power of Hammer

The following table, which gives the maximum diameter of ingot that each power of press is capable of dealing with effectively and the equivalent power of steam-hammer, will be of assistance in making a comparison of the two methods of forging:—

Ingot Diameter		Press Power tons	Hammer Power tons
In.	mm.		
5	125	100	0.50
6	150	150	0.75
8	200	200	1.00
10	250	300	2.00
12	300	400	3.00
14	350	500	4.00
16	400	800	5.00
20	500	800	7.00
24	600	1000	10.00
30	750	1200	15.00
36	900	1500	20.00
48	1200	2000	40.00
60	1500	3000	80.00
72	1800	4000	120.00
84	2100	5000	
96	2400	6000	

From this table it will be seen that the power of the press bears a fairly direct proportion to the diameter of the ingot, but the power of the hammer required increases much more rapidly, being nearly proportional to the square of the diameter or the sectional area of the ingot. For instance, for an ingot 24 in. (600 mm.) in diameter, the power of press is 1,000 tons, and the hammer 10 tons; but for a 48-in (1,200 mm.) ingot, the power of press is 2,000 tons, and the corresponding hammer 40 tons, and. consequently, the heavier the work the greater is the advantage of the press over the hammer.

As stated above, for the heavier classes of work, presses have been used in many of the principal steel works for as long as thirty years, the power of such presses most generally used, up to twenty years ago, being from 2,000 to 3,000 tons, though during this period some much more powerful presses, up to 10,000 tons, have been put down for forging armor plates and for special purposes. During the next ten years—that is, up to ten years ago—owing to the increased size of guns and other forgings required, a good many presses of 4,000 tons have been adopted.

During the last ten years—that is up to the present date—owing to the further increase in the size of guns, and also to the larger forgings required for marine work, particularly turbine-drums and spindles, the power of press required has increased correspondingly, 6,000 tons being the power most recently adopted for the heaviest work of this class.

Method of Driving Press

To explain the development in the use of the press, some mention must be made of the improvements that have been introduced in the method of driving. The earlier forging-presses have generally been driven by means of pumping-engines working direct into the press cylinders, the working pressure being usually from 2½ to 3 tons per sq. in. (400 km. to 480 km. per sq. cm.). For large presses this method gives satisfactory results, but it has never been used to any extent for presses of medium or small power. for which greater simplicity and a higher speed of working, especially in finishing, are more essential.

Another method is to make the press purely hydraulic, working it from an accumulator. For certain special work, such as piercing projectiles or billets for tube-making, in which a long stroke without any pause is desirable. this system is undoubtedly the best. For ordinary forging operations, however, for which a short, rapid stroke is essential, this system has only been adopted to a limited extent and is applicable only to presses of comparatively small power.

The system that .fulfils the requirements of forging best .is the steam, hydraulic intensifier. For a good many years it has been used to some extent, but it is only within the last ten years that it has come into general use and

been adopted extensively for presses of small and medium power as well as for the heaviest forging-presses. The advantages of the steam-intensifier presses are that, in its latest and best form, it is capable of fulfilling adequately all the requirements of forging.

These include sufficient rapidity of action in all its movements—that is, the idle stroke when lifting the press-head and bringing it down to its work and the forging stroke or penetration. The idle stroke should be from 6 in: to 12

FIG. 2. ELECTRICALLY-DRIVEN TURN-
ING GEAR FOR ROTATING
FORGINGS.

in. (152 mm. to 305 mm.) a second, according to the power of the press and the nature of the work, and the forging stroke up to 3 in. (76 mm.) per second, according to the resistance of the forging. Also there should be no pause or dwell either when the tool first touches the work—that is, before the forging stroke commences— or at the end of the forging stroke; that is to say, the return stroke should commence directly the penetration is completed, any pause causing loss of time as well as chilling the forging. The length of the forging stroke is usually from 1 in. (25 mm.) up to 6 in. (150 mm.), or 8 in. (200 mm.) in

the case of very large presses and the speed of such forging from 10 up to 50 strokes a minute, a total penetration of 60 in. (1,500 mm.) a minute being obtainable when the power of the press is well up to its work and the forging handled expeditiously.

For rounding or finishing a forging, a very rapid stroke is desirable; a good intensifier press of as much as 6,000 tons being capable of working up to 40 strokes a minute, and smaller presses proportionately quicker up. to 100 strokes a minute. With these speeds of working an efficient control-gear is desirable both to limit the forging stroke and also to prevent any overrunning of the intensifier in case the load on the press is suddenly removed, owing to the forging or tool slipping or other accidental cause. So successfully has efficient control-gear been applied to intensifier presses that they can be worked quite safely and easily at the high speeds mentioned. Presses already constructed on this system range from 100 up to 6,000 tons power, the larger presses giving proportionally equally as good results both in forging and in rapid finishing strokes.

Forging Operations

Fig. 1 illustrates an up-to-date forging-press of 6,000 tons power, engaged in forging a solid shaft. This press is driven by two intensifiers, arranged to work simultaneously or independently by means of catches on the handing-lever. Working simultaneously, the intensifiers give a forging stroke of 10 in. (250 mm.) and a speed of from six to ten strokes a minute, according to the penetration required. When short strokes for finishing are required, either intensifier can be used and the press can be worked up to forty strokes per minute.

The full equipment for such a press, to enable it to deal advantageously with the various classes of work, includes the following:—

Mandrel-gear, giving the mandrel-blocks a travel of about 30 ft. (9 m.) on each side of the press.

An extended base of sufficient strength so that the full power of the press can be exerted on the mandrel-blocks when they are spaced at the maximum distance apart required for expanding work. such as turbine-drums or gun-jackets.

Manipulating .gear for handling forgings without the use of a crane:—this is particularly useful for certain classes of work, such. as forging armor-plate and turbine-wheels, and also in some cases for cogging ingots.

Transverse tool-changing gear:—this is useful for forging crank shafts and similar work when it is desirable to change from flat tools to V-shaped or swaging tools. The gear enables both top and bottom tools to be changed very quickly without taking the forging from under the press.

Turning-gear for rotating the forging:—The most convenient form is a self-contained electrically-driven gear, suspended from the crane-hook. Fig. 2 illustrates such a gear. Preferably this should be provided with a friction-clutch which slips during the instant that the press grips the forging.

A press with the above equipment is suitable for practically all classes of heavy work. Examples of forgings made under such a press will be given in the continuation of the present article next week.

———————

OPEN-HEARTH VS. ELECTRIC FURNACE FOR COMMERCIAL STEELS

THE fundamental question is one of cost, and any authentic figures are always welcomed as throwing fresh light on this subject, which is of great commercial interest. The fact that the electric furnace can successfully compete with the crucible process, and in some cases with the small converter, is, we think, established, but a comparison of the electric furnace with modern open-hearth furnace is open to discussion, and is made the subject of an article by S. Cornell in a recent issue of the Metallurgical and Chemical Engineering.

In order to provide data on this point a table has been compiled for a year's operation of a large open-hearth plant and compared with what is claimed to be practice obtainable in an electric furnace plant of the same capacity. The open-hearth plant consisted of 80-ton furnaces, having a production of 200 tons of ingots per day for each furnace, compared with 20-ton electric furnaces, approximately the same capacity per day.

Dealing with the materials necessary to make one ton of steel, and including all consumable material and cost of repairs, it is calculated that the cost per ton of steel is $14.50 for the open-hearth against $18 for the electric steel. The cost and quantity of raw material is taken to be the same in both cases, and the difference is due entirely to a fuel cost for producer gas being 54 cents for the open-hearth, as against $4.40 for electric power in the electric furnace.

The general labor charges which are common to both processes is worked out at $1 per ton of steel, whilst the cost of general repairs is given as 6 cents per ton.

From a careful consideration of the outlay necessary in installing electric furnaces of sufficient capacity to produce the same tonnage as a set of open-hearth furnaces, it is calculated that in order to compete in cost the electrical

energy must be produced at below 1 cent per kilowatt-hour.

The present electrical equipment using blast furnace gas cannot do better than 7 cents per kilowatt-hour.

———————

HINTS TO FOUNDRYMEN

A CONTRIBUTOR to "Engineering" says it appears to be general practice to make patterns for all classes of castings, light or heavy, and to ignore loam work as being either too costly or causing too much dirt in the shops. This is entirely wrong, as I think some of those in charge have but little knowledge of how loam work can be made to advantage over sand, both as regards cost and keeping the shop just as clean.

I was talking to a gentleman the other day, who thought that his was quite a modern concern. They were engaged on one class of engine, etc., or what might be called repetition work. There was one casting in particular being made from patterns; this was what molders commonly call a block fly-wheel,

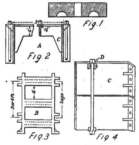

ILLUSTRATING HINTS TO FOUNDRYMEN.

up to 12 ft. in diameter, as shown in sketch (Fig. 1). Now to make such castings as these in sand is not only increasing the cost, but causing a lot of unnecessary hand labor for the molders.

Anyone that claims to know anything of modern foundry practice could take 10 per cent. from the cost of a casting of this kind by making it in loam, and more than another 10 per cent. from all others that were required. There is plenty of room yet for green, dry sand and loam work in a modern foundry, and castings can only be produced economically by sound judgment in the use of them.

Jobbing Foundry-Box Parts

The writer has worked in plenty of jobbing foundries doing a fairly decent class of work that could be cast on its joint, but if anything came along that had to be cast vertical, it had to be rejected as being something out of their particular line. Now if the boxes had

been made to tackle any class of work, such castings as hydraulic bodies, rams, etc., could have been made.

All that is required is a crane of sufficient height to turn them up, to make the bars in the box with lips as shown in the sketch (Fig. 2) like a T, and to place lugs (Fig. 3) at the back about 3 ft. or 4 ft. apart, for the boxes to be bolted with these when they are being used vertically. A cross-section of the box is shown at A; the top and distance of the bars should be apart are indicated at B, while C shows a side view of a pair of boxes, and D the bolts and method of fastening. These can be made in lengths to the requirements of the foundry; 6 ft. and 9 ft. are fairly handy. The boxes can be bolted at the ends to make any length required; they are perfectly safe, no plating or the back of the bars or ramming being required.

Splitting Castings

The best method of treating splitting plates for castings is to glue a thin layer of asbestos over them; when cast they will rap out, leaving a smooth, clean face.

———————

LARGE SHELLS: PRODUCTION PROBLEMS AND POSSIBILITIES.—V.

(Continued from page 545.)

meter of the inside of the pulley, and is of the taper cork insert type. It is operated by a separate valve and the release is by springs. The thrust of the spindle is taken at the forward end bearing, a take-up collar being provided at the rear of the same bearing.

The tool base is bolted firmly to the bed and carries both roughing and finishing heads. On the 9.2 in. the roughing is split into two operations. Each tool is fully adjustable, but when once set, they can be removed and replaced with precision. The feed of both tools is by a single screw to positive stops. The undercutting tool is on a separate block mounted on the front slide at the correct angle and is fully adjustable. The feed is through lever and cam. The rear or finishing tool is so mounted that the tool slides past the band and shears it to the exact size and shape. The tool block is securely bolted to the base, but is slightly adjustable to and from the cut. It can also be rotated if necessary, so the tool will cut both ends of the band to correct size. The feed is by rack and pinion, with handle in a convenient position for the operator.

Two roller supports are provided on the bed so that the shell can be lowered by an air hoist or other means into the chuck, and in line with the centre. The rollers are not in contact with the shell when it is rotating.

EDITORIAL CORRESPONDENCE

Embracing the Further Discussion of Previously Published Articles, Inquiries for General Information, Observations and Suggestions. Your Co-operation is Invited

GALVANIZING CONDUITS

THE manufacture of galvanized conduits at the plant of the Orpen Conduit Co., Toronto, Ont., presents several interesting features, the principal one being that the pipe is copper-plated before being galvanized. Copper-plating is a protective covering of great value in resisting corrosion, forming an ideal "couple" between the metal of the tube and the zinc. The zinc is, therefore, deposited on copper and not on iron by means of the electro-deposition process, thus its full value as a rust preventative is secured.

This conduit, which bears the trade name of "Xceladuct," is a high-grade mild steel butt-weld pipe of special quality, being heavier than ordinary tubing. The pipe for the straight conduit is received at the factory in 10-ft. lengths, and is made in Canada; the bends are, of course, shorter. The pipe is threaded at both ends, one end having a coupling when shipped. The conduit is made in sizes from ½ in. to 6 in. inclusive.

Pickling Process

The first process consists of pickling, which is done to remove the scale, etc., from the pipe. For this process there are installed a number of wood tanks, the first containing a potash solution for removing oil or grease from the pipe. The pipes are then dipped in a tank containing sulphuric acid, which removes the scale. They are then dipped in another tank containing muriatic acid, which gives a smooth surface to the pipe.

On the completion of the foregoing, the pipes are laid out on a flat table, where they are carefully inspected. A powerful light at one end enables an examination of the inside to be made. Any defective pipes are laid to one side, and the sound ones are taken over to the plating department.

Plating and Galvanizing

In the plating department are installed a series of five tanks in one row, each being 14 feet deep, and having a capacity of 4,000 gallons. The tanks sit in an asphalt pit, and are constructed of wood or steel, according to the character of solution which they contain. Before being dipped in the first tank, the pipes are assembled in a "basket," in which they are held until this process is completed. The basket is made of copper, and is constructed so as to hold the pipes vertically. It has a capacity for 175 lengths, each 10 ft. long, of ½-in.

pipe, equivalent to 1,750 ft. The larger sizes, being heavier, fewer pipes are dipped at one operation.

The basket is suspended from and carried along by means of an electrically-operated hoist on an overhead runway, extending the full length of the tank. At the end of the operation, when empty, the baskets are returned on an overhead runway at each side of the tank. The hoist has lifting and lowering motions for use when dipping the pipes.

The pipes are dipped in three different solutions before being copper-plated. After being in the copper-plating tank the prescribed length of time, they are taken out and dipped in the zinc tank, where they remain about 45 minutes. They are then taken out and put in the rinsing tank.

The electrical equipment for the zinc and copper tanks consists of two motor-driven plating dynamos. One unit consists of a Canadian Hanson & Van Winkle Co. plating dynamo, 8,000 amperes, 6 volts., direct-connected to an 85 h.p. C.G.E. motor running at 470 r.p.m. The other unit is of the same make and size, but belt-driven by an 80 h.p. C.G.E. motor. The wiring from the dynamo is taken under the floor, up the outside of the tank, and connected to the anodes.

Drying and Insulating

After the pipes have been rinsed, they are taken from the basket and placed vertically over a grating through which air is passing from a blower. The pipes are thus thoroughly dried before being insulated and again inspected. For the insulating or enameling process, the pipes are laid inclined on a table and the enamel flows from an overhead tank through them into a shallow tank, whence it is pumped back by means of a motor-driven rotary pump to the overhead tank. The pipes are then placed vertically over a grating to dry, a current of air passing through accomplishes this.

They next go to the labeling table, where they are laid flat and two labels affixed—one factory label and one for the Underwriters. At this stage the pipes are carefully examined by the Underwriters' inspector, who selects one from each batch to be tested. Before leaving the table, a coupling is screwed on one end of each pipe, after which they are tied in bundles ready for shipping. The test pipe is cut in two sections on a hack saw machine, and a length taken to the inspector's office.

where it is tested to ascertain if the zinc coating and enamel are adhering in a satisfactory manner. The tests are very thorough, as the conduit must conform in all respects to Underwriters' rules.

Elbows

The above-mentioned processes cover the straight lengths of conduit only. The elbows are treated in a somewhat indifferent way, chiefly as regards handling. Short lengths of pipe are received at the factory, threaded at both ends. The first requirement is to bend them to the required angle and radius. For the larger pipes a belt-driven bending machine operated from the line shaft is used, while the smaller size pipes are bent in a steam-operated machine. In both cases different forms are used to suit the various sizes of pipe.

The elbows, as in the case of the straight pipes, undergo the pickling process and in the same set of tanks. The elbows are, however, scrubbed with sand after being pickled. They are then copper-plated in the same tank as the straight lengths, but have a separate tank for the galvanizing process. The current for the zinc tank is furnished by a plating dynamo, which is also connected to the zinc tumbling tank used for galvanizing the couplings. The dynamo supplies current at 6 volts, 400 amperes. After being galvanized, which takes about 30 minutes, the elbows are taken out and hung on a rack to dry, and then taken over to the insulating tank to be enameled inside. After being enameled, they are hung on racks to dry, and are then inspected and stamped the same as the straight conduit. The elbows are also inspected and tested between the various processes. All sizes are treated in the same way.

Couplings

As already stated, a coupling is supplied with each length of conduit. The first process consists of pickling, which is done in the same set of tanks as the pipe and elbows. They are also copper-plated in the same tank, and are held in wire baskets during the process. For galvanizing, there is a special tumbling tank installed, in which the couplings are revolved for three hours. When the galvanizing process is completed, the couplings are taken out and washed in cold water. They are then put in wire baskets, washed again in hot water, and afterwards laid on a wire screen to dry. They are afterwards taken to the store room to be used as required.

CHUCKS VERSUS FIXTURES
By D. A. Hampson.

ALMOST every part that is made in quantities of one hundred or more requires some sort of fixtures if the work is to be done cheaply and the parts have to be at all interchangeable. The commercial side should determine how elaborate these fixtures are to be; but before going to the expense of making or even of designing them, it is well to take account of stock and see if there are no standard adjustable fixtures which can be used instead.

Into this class come chucks,

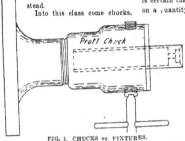

FIG. 1. CHUCKS vs. FIXTURES.

dull heads, jigs, vises, etc. On work that is to last but a few days or a few weeks, such tools can be made to show almost as great a production as specially made-up tools, and show a nice saving in initial expense and ultimate discarding. Two cases in which ordinary drill chucks were used to advantage will be reviewed.

Ten thousand brass shells were to have an ⅛-inch hole drilled through the end. Instead of making a special holder for these, a Pratt drill chuck of the familiar two-jawed type was selected from among the "spares" and a cast iron base made to support it. The drawing shows this and also the wrench for opening and closing, the wrench being made a tight fit in the socket, so it is always in place. A quicker or more satisfactory

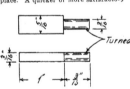

FIG. 2. CHUCKS vs. FIXTURES.

fixture would have been impossible to make, and when the job was done we had not added as much as a dollar to the obsolete tool account.

A "trial lot" of a new mechanics' tool required a thousand pieces of 3-16 x 5-16 drawn steel, with a symmetrically turned stem at one end. There was in the tool room a nearly new two-inch, two-jawed Cushman drill chuck. To this we fitted a shank that was fitted to the "centre" taper of a lathe spindle. The jaws of the chuck being V-shaped, formed a perfect driving and centering medium. After milling the pieces to length, they were turned in the chuck, suitable stops having been arranged for position, length and diameter. It is doubtful if any special fixture would have done the work more accurately—it is certain that they would not have paid on a quantity of this size.

EXPANDING MANDREL
By G. Barrett

THE expanding mandrel is undoubtedly one of the most called-for tools required in the manufacture of shells, and its use has never been so widespread as it is at the present time. The one which we illustrate herewith was designed for a lathe which had been temporarily out of commission owing to the lack of a hollow spindle.

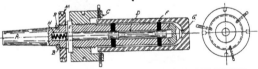

AN EXPANDING MANDREL.

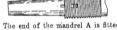

The end of the mandrel A is fitted to the tapered hole in the lathe spindle and is prevented from turning by means

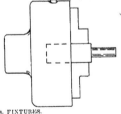

of the pins B which engage with two corresponding slots in the spindle nose. In operation the shell is first placed over the mandrel till it touches the point G. The nut C is then tightened against the cotter F, causing it to draw up the rod D, the inner end of which projects into the hole H. By this means the tapered diameters on the rod D force the dogs F outward against the shell. By tightening the three set-screws shown in the nut, a firm grip is assured, since the strain of the cut will have the tendency to still further tighten the nut.

To remove the shell the set-screws are slackened off and the nut C is screwed back. A slight pressure is then applied to the cotter E by means of the nut M. This forces the rod back and relieves the strain on the dogs sufficiently to allow the spring H to force the rod back the remaining distance and thus release the shell.

TRADE WITH BRITAIN

AN increase in Canadian trade with Great Britain, particularly in exports, is shown in British trade returns for the nine months of the present fiscal year ending October 31st last, which have reached the Trade and Commerce Department. The 1915 figures of Canadian trade for this three-quarter period were $188,824,340, as compared with $174,000,000 in 1914. Of the former total, over $138,500,000 was in Canadian exports. Canada's trade was considerably less than Australia's, which totalled $300,000,000, probably due to the purchase of war munitions. Great Britain's trade with the United States increased by about $350,000,000. The October figures also show a decrease in the disproportion evident between British imports and exports since the war. From this, the inference is drawn that the English people are beginning to economize by cutting down consumption of all but necessities.

Truing an Oilstone.—To true an oilstone, the Mechanical World suggests the following method:—Take a piece of soft pine board of any thickness, about 8 in. wide and 3 ft. or 4 ft. long. Lay it on a bench and fasten it with a hand-screw or other clamp. Put on some clean, sharp sand, screened about as fine as that used for plaster work. Use no water, and rub the stone back and forth over the board in sand. This will give a flat surface to the stone in a short time. Care should be taken to move the stone on straight lines, so as not to give it a warped surface. If a fine surface is wanted, a finer grade of sand or sandpaper may be used to finish with.

PROGRESS IN NEW EQUIPMENT

A Record of New and Improved Machinery and Accessories for the Machine, Pattern, Boiler and Blacksmith Shops, Planing Mill, Foundry and Power Plant

GRINDING ENDS OP SHELLS

A T the square end of a shrapnel shell there is a small turning hub or centre projection which has to be removed before the shell is completed. There are various ways of removing this stock, but production is the essential factor in all. The Gardner Machine Co., Beloit, Wis., has developed a special grinding machine for this operation,

This equipment is also used for cutting off the square or angular hubs from base plates in high explosive shells as well as on shrapnel casings. From a

ings. On each end of the spindle is mounted a "Perfection" ring wheel chuck, of improved design. These chucks contain 16-in. diameter abrasive rings selected for these operations.

Lever feed work tables, supported on a rigid rocker-shaft running entirely through the base, serve the grinding wheels. The shell is quickly clamped into "V" fixture, and by means of a hand lever is firmly forced against the grinding wheel up to a micrometer stop, and at the same time rocked across its face.

A pump and water system driven

lever feed work tables, two shell-holding fixtures, water pump, with connections, countershaft and general supplies. The whole outfit weighs 4,000 pounds.

This firm also builds a No. 50 grinder, which is much heavier and more powerful than the one illustrated herewith, and which is also being adopted by manufacturers of 6-in. shells for these same operations.

MACHINE FOR GRINDING BASES OF SHRAPNEL SHELLS.

grinding standpoint, the operation is the same in both cases. In some instances the hub is removed by some other process and the riveting done. It is then placed on the grinder and the balance of the base plate is removed, taking a light cut over the entire base of the shell as well.

The machine consists in the main of a very heavy base casting, with a 2-in. diameter spindle, mounted in either high-grade babbitt bearings or ball bear-

from the counter-shaft directs the water or grinding compound at the point of grinding contact. The hoods and basin are one-piece castings.

The output obtained from one end of the machine varies according to the size of the projection to be removed, being from 40 to 100 per hour. This machine, with both ends equipped and operated by two men, gives double this output.

The complete equipment includes two 16 in. "Perfection" chucks, two 16 in. abrasive ring wheels, two semi-universal

DIAL FEED ATTACHMENT FOR LARGE SHELLS

THE Ferracute Machine Co. of Bridgeton, N.J., has recently developed an automatic dial feed attachment for use in cupping and re-drawing cartridge cases and other sheet-metal shells, the attachment being used in connection with straight-sided presses exerting pressures from 100 to 200 tons.

The illustration shows a 100-ton press fitted with one of these attachments, the dial having six recesses, it being essential that a recesss be in alignment with the centre of ram as it descends to insure the entrance of the punch in the partially drawn shell. Pressure on the treadle will cause the press to stop instantly at any portion of its stroke.

The dial has an intermittent motion, being at rest while the press makes a stroke. A cam on the main shaft gives motion to the rack that partially revolves the dial yoke. An additional cam on the main shaft operates the "lock" mechanism. This lock is a lever containing a wedge-shaped end that fits each of the six notches in the circumference of the dial.

Connected to the lock mechanism is an "interrupter." When the lock is in a notch, indicating that the ram and dial-recess are properly aligned, the interrupter prevents a weighted rod from descending, but if by any mischance the lock should fail to enter its notch, the interrupter will change its position, allowing the weighted rod to descend and depress the treadle, causing the press to instantly stop, and thereby avert damage. After cause of trouble has been ascertained and adjustment of dial finished, the press may be made to complete its stroke by pulling down the handle-strap, the ascending ram elevating the weighted rod, and the treadle and treadle-rod assuming their first position.

The press is equipped with a combined friction clutch and brake. Pulling down the hand-lever connects the power to the shaft, giving motion to the press,

while depressing the treadle releases the clutch and applies the brake. The machine is, therefore, under perfect control at any position of its stroke.

There are several adjustments that give accuracy to the motion of the dial. For instance, the amount of rotation of the dial may be regulated by means of a roller at top of yoke, set-screws being provided to give exact position. The motion is also controlled by a brake which is automatically released during the locking process. There is an eccentric adjustment which provides for the accurate relation between the lock and its notch, the tension of the lock being regulated by means of an adjusting spring.

Although designed primarily for continuous action, the press with its attachment can be used intermittently, that is, automatically stopping at the end of each stroke. When running continuously, 12 shells per minute are produced.

———— ❀ ————

1,000-TON HYDRAULIC CARTRIDGE CASE HEADING PRESS

THE hydraulic press, illustrated, was designed for the heading of brass shells after they have been indented. The heading operation is accomplished by inserting a fullering block between the head of the press and the top of the cartridge case, the latter being held in place by a suitable die. As the pressure is applied, the fullering block causes the brass to flow outward

in all directions, thus forming the head of the shell.

The press has a revolving turret with dies to receive three shells. This provides for an almost continuous operation, as there is always one shell awaiting the heading operation and one shell being unloaded, while the other shell is undergoing the heading operation. The rotation of this turret is controlled by an indexing device, so that the shell is accurately held in place directly beneath the fullering block.

The indexing device is operated by a lever conveniently placed for the operator. By merely pulling the lever, the index latch is disengaged from the turret, clearing and releasing the turret for revolving. The photograph shows a rear view of the press. The lever controlling the indexing device is shown to the extreme right. The operating valve is entirely hidden by the press.

The turret revolves on a spindle mounted in ball bearings, which are set in the front side of the press between the strain rods. Handles placed around its outer edge at an equal distance apart are fitted for the purpose.

After the heading operation is completed, the shell is pulled from the die by an hydraulic ejecting device, which consists of two rams, one attached to the head of the press and the other to the base. Spring gripper jaws are attached to the upper ram for gripping the head of the shell. Each of these rams is provided with two auxiliary pull-back rams.

DIAL FEED ATTACHMENT FOR CUPPING AND RE-DRAWING CARTRIDGE CASES.

1,000-TON HYDRAULIC CARTRIDGE CASE HEADING PRESS.

The ram attached to the base of the press forces the shell up out of the die to a point where the gripper jaws on the upper ram can readily take hold of the newly-formed head. Pressure is then applied on the auxiliary rams for the return of the ejector rams. The finished shell is next removed from the gripper jaws of the upper ram and the device is ready for the next shell.

The use of two ejector rams, each working toward a common centre, makes the ejecting operation much more rapid and clears the turret in less time, so that it can be revolved more quickly for heading the next shell. The maximum pressure of the two ejector rams is 5 tons each.

The press is of steel construction throughout, the base and cylinder being cast in one piece. The ram has a diameter of 28 inches and a run of 12 inches. The press has an adjustable daylight of 30 to 45 inches.

The maximum total pressure capacity of the press is 1,000 tons, and the number of pressings per minute depends upon the capacity of the pump used for its operation. As a basis, the press will make 3½ pressings per minute with the pump furnishing 19¼ gallons of water per minute at a pressure of 3,250 pounds per square inch.

This projectile press is a new design and recent addition to those being built by the Hydraulic Press Mfg. Co., Mount Gilead, Ohio.

DOMINION ECONOMIC COMMISSION

AS a first step of the recently-appointed Federal Economic Commission, under Hon. J. A. Lougheed, towards securing working data from which to draw conclusions, a special census of Canadian industrial companies is to be taken. At the conclusion of the meeting held on Dec. 13 of the commission, R. H. Coates, Director of the Census and Statistics Branch, announced that a census of manufacturers would be taken by post next month. All manufacturers will be asked to fill out a census form, giving details as to capital invested, number of employees, wages, etc., and volume of production, distinguishing between war orders and general business.

Efforts also will be made by the Commission to get the latest comprehensive and reliable information regarding agricultural production, including cost of production, market facilities, etc. In this connection the various agricultural associations will be asked to co-operate.

After the preliminary information sought is received, the Commission will visit various centres in sections and take public evidence as to needs, and re-

commendations looking to a better conservation of the national resources, and more economic methods of marketing, etc.

HIGH-DUTY POWER PRESS

FORCED production of automobile parts is creating a demand for stronger and heavier presses, with greatly increased factors of safety. The machine illustrated was recently developed by the Cleveland Machine & Mfg. Co., Cleveland, Ohio, for the production of automobile frame cross-bars and similar heavy forming, blanking and drawing.

A heavy spring drawing attachment (not shown) is mounted in bed of press, and is used for operating knock-out pads

HIGH DUTY POWER PRESS.

and drawing or pressure rings. The gearing is of steel throughout, with machine-cut teeth. The friction clutch is multiple disc type and very powerful.

The principal dimensions of the machine are as follows: Gross weight, 167,000 pounds; bed area, 72 in. R to **L**, x 48 in. F to B.; crank shaft, 10 in. diameter in bearings and 11¼ in. on crank pins; stroke of slide, 10 in.; distance bed to slide, stroke down and adjustment up, 26 in.; proportion of gearing, 45:1.

QUICK-FIRING GUNS

THE rapid development of artillery during the last few years has naturally led to the coining of a number of phrases to express the difference between the various types, and distinguish the new from the old and obsolete; while the rapid education of the general public in the elementary principles of modern warfare has brought some of these phrases into common use. It not infrequently happens that a whole series of remarkable improvements are designated for

general use by a single expression. An excellent example of this is the term "quick-firing" as applied to modern gunnery.

Less than twenty years ago, it was recognized that the gradual development of the field gun had made so many changes necessary in the tactics of infantry that a radical improvement in the speed and accuracy of artillery fire was desirable. Men were no longer fighting in masses in the open, and they had found it more effective to offer only a fleeting target to the gun. Consequently the time required to load, run out, and aim the gun had to be reduced if the weapon was going to be of much use. Loading was enormously accelerated by the adoption of the single-acting breach mechanism, combined with a device for ejecting the empty cartridge case. The running in of the old gun to the firing position was necessary, because the recoil after firing was taken up by the gun and carriage as a whole. Hence, the weapon had to be run forward, set in position and re-aimed.

The Modern Gun

In the modern gun the barrel only moves, the energy of the recoil is taken up in an elastic press, and the barrel is returned to its original position by a spring. After firing, the modern gun returns instantly to the position it occupied before firing, and generally this does away with re-aiming. The improvements enumerated were first adopted by the French in 1897, and their famous 75-mm. gun can fire from twenty to twenty-five rounds per minute. Germany carried out an elaborate series of trials, extending over a long period, but it was only in 1905 that they adopted the quick-firing gun.

In the services the Maxims, Gatlings, and other automatic feed guns are not spoken of as quick-firers, that name being given to all weapons which are loaded by hand. Thus the 6-in. naval gun is a quick-firer and represents the maximum size of a quick-firing gun in the navy. The shell weighs 100 lbs., and can just be man-handled. The 7.5 in., the 9 in. and all heavier weapons are provided with regular hydraulic hoists and loading arrangements.

At the other end of the scale is the little 3-pounder, and from the 6-in. to the 3-pounder the firing arrangements are sensibly the same. Naval guns of greater than 6-in. calibre differ in the first instance from the 6-in. gun, in that the loading is done by machinery. With the main armament weapons of our battleships, the size and weight of the breech blocks renders it necessary to open and close the breech by hydraulic power; but the principle of the recoil arrangements is the same throughout.—Liverpool Journal of Commerce.

The MacLean Publishing Company
LIMITED
(ESTABLISHED 1888)

JOHN BAYNE MACLEAN - - - - - President
H. T. HUNTER - - - - - General Manager
H. V. TYRRELL - - - - Asst. General Manager

PUBLISHERS OF

CANADIAN MACHINERY
AND MANUFACTURING NEWS

A weekly newspaper devoted to the machinery and manufacturing interests.

PETER BAIN, M.E., Editor.

Associate Editors.
A. G. WEBSTER. . J. B. WILSON, J. H. RODGERS.

B. G. NEWTON - - - - - Advertising Manager

OFFICES:

CANADA—
Montreal—Rooms 701-702 Eastern Townships Bank Building.
Telephone Main 1255.
Toronto—143-153 University Ave. Telephone Main 7324.
UNITED STATES—
New York—R. B. Huestis, 115 Broadway. Phone 8971 Rector.
Chicago—A. H. Byrne, Room 607, 140 South Dearborn St.
Telephone Randolph 3234.
Boston—C. L. Morton, Room 733, Old South Bldg.
Telephone Main 1024.
GREAT BRITAIN—
London—The MacLean Company of Great Britain, Limited, 88 Fleet Street, E.C. E. J. Dodd, Director. Telephone Central 12960. Address: Atabek, London, England.

SUBSCRIPTION RATES:

Canada, $2.00; United States, $2.50; Great Britain, Australia and other Colonies, 8s. 6d. per year; other countries, $3.00. Advertising rates on request.

Subscribers who are not receiving their paper regularly will confer a favor by letting us know. We should be notified at once of any change in address, giving both old and new.

Vol. XIV.　　DECEMBER 16, 1915　　No. 25

PRINCIPAL CONTENTS.

PREPARING FOR COMMERCIAL ASCENDENCY

THE question of Canadian manufacturers being able to supply to the fullest extent possible, immediately, the munitions demand subsides, our own domestic requirements and develop concurrently and maintain an export trade that will transcend in volume all previous achievement, is one that should be considered and given tangible expression to now by individual or collective organization. We have organized for munitions production in a manner that has merited world-wide commendation. We can also organize for these others just as effectively, and without hampering in the slightest degree the meantime munitions production efficiency.

Organization for the manufacture of munitions was largely an individual firm affair, and, the Department of Trade and Commerce, Ottawa, notwithstanding, similar action is inevitable if we would corall in large part our domestic business and establish a world-embracing export trade. Evidence is not lacking that preparation is proceeding apace in the United States and even in Great Britain to not only maintain but to take hold of every available opportunity, both domestic and foreign.

In the United States, the American International Corporation has been organized with a capital of $50,000,000, and with the most prominent of that country's financiers and business specialists behind it. The avowed intention is to prosecute the most vigorous foreign trade expansion campaign in American history.

In Great Britain, irrespective altogether of the duration of the war, it is expected that in a very short time with the completion of the many large and specially-equipped munitions factories, there will be a wholesale release of firms, large and small, meantime engaged in the production of war material. By way of anticipating this, the Imperial Government, through the Prime Minister and Chancellor of the Exchequer, has stated definitely that the maintenance of a large and increasing export business is one of the essentials if victory is to be complete.

Export trade on Britain's part is, even now, war work, and the immediate future is ear-marked for its diligent prosecution. By no other means can her resources be husbanded and her efforts be so thoroughly concentrated on the task she and her Allies have undertaken to bring to its logical conclusion than that of retaining the world-wide commerce and market for her wares already established, and by enlarging their scope and service.

We have already hinted at the almost certainty of Canada's efforts lacking official Governmental backing. Under such circumstances individual firm enterprise must needs be forthcoming. Efficiency engineers, safety experts, and a myriad other departmental appointments have come to be recognized as part and parcel of our leading metal-working and general manufacturing establishments. May we suggest that a domestic and foreign trade department is now also highly desirable, manned by experts in each particular sphere, who will organize campaign plans, collect data, and generally arrange easy transition from war to peace-time pursuits.

Men fitted to grapple successfully with the problems of shell-making were neither hard to "spot" nor to develop, and we are of opinion that their prototypes having both a sufficiently wide commercial experience and intimacy with the manufacturing or practical side of any factory product, will be likewise as promptly available and as prolific of achievement.

Our plants generally are in a position with the cessation of hostilities to compass a greatly increased and more diversified output. To secure business in the coming time, to maintain full employment for men and machines will, however, necessitate "rustling" for both home and foreign orders. It should also be realized that war orders have increased the capacity and diversity of output of our competitors, making former competition none the less, but rather a great deal more. The need of organizing is therefore particularly and peculiarly urgent.

SELECTED MARKET QUOTATIONS

Being a record of prices current on raw and finished material entering into the manufacture of mechanical and general engineering products.

PIG IRON.

Grey forge, Pittsburgh	$17 95	
Lake Superior, charcoal, Chicago	17 75	
Ferro nickel pig iron (Soo)	25 00	
	Montreal.	**Toronto.**
Middlesboro, No. 3	$24 00	.....
Carron, special	25 00	.....
Carron, soft	25 00	.....
Cleveland, No. 3	24 00	.....
Clarence, No. 3	24 50	.....
Glengarnock	28 00	.....
Summerlee, No. 1	30 00	.....
Summerlee, No. 3	29 00	.....
Michigan charcoal iron	28 00	.....
Victoria, No. 1	24 00	24 00
Victoria, No. 2X	23 00	24 00
Victoria, No. 2 plain	23 00	24 00
Hamilton, No. 1	23 00	24 00
Hamilton, No. 2	23 00	24 00

FINISHED IRON AND STEEL.

Per Pound to Large Buyers.	Cents.
Common bar iron, f.o.b., Toronto	2.50
Steel bars, f.o.b., Toronto	2.75
Common bar iron, f.o.b., Montreal	2.50
Steel bars, f.o.b., Montreal	2.75
Twisted reinforcing bars	2.55
Bessemer rails, heavy, at mill	1.25
Steel bars, Pittsburgh	...
Tank plates, Pittsburgh	...
Beams and angles, Pittsburgh	...
Steel hoops, Pittsburgh	
F.O.B., Toronto Warehouse.	**Cents.**
Steel bars	2.75
Small shapes	2.75
Warehouse, Freight and Duty to Pay.	**Cents.**
Steel bars	2.20
Structural shapes	2.30
Plates	2.30

Freight, Pittsburgh to Toronto.

18.9 cents carload; 22.1 cents less carload.

BOILER PLATES.

	Montreal	Toronto
Plates, ¼ to ½ in., 100 lb.	$2 75	$2 75
Heads, per 700 lb.	3 00	3 00
Tank plates, 3-16 in.	3 00	3 00

OLD MATERIAL.

Dealers' Buying Prices.	Montreal.	Toronto.
Copper, light	$13 75	13 50
Copper, crucible	16 25	16 00
Copper, unch-bled, heavy	15 75	15 00
Copper, wire, unch-bled	15 75	15 25
No. 1 machine compos'n	12 00	11 75
No. 1 compos'n turnings	11 00	10 00
No. 1 wrought iron	10 00	10 00
Heavy melting steel	9 50	9 50
No. 1 machin'y cast iron	13 50	13 00
New brass clippings	11 50	11 00
No. 1 brass turnings	9 50	9 00
Aluminum	29 00	29 00
Heavy lead	5 25	5 00
Tea lead	$4 25	$4 25
Scrap zinc	12 75	12 00

W. I. PIPE DISCOUNTS.

Following are Toronto jobbers' discounts on pipe in effect Nov. 5. 1915:

	Buttweld Black	Buttweld Gal.	Lapweld Black	Lapweld Gal.
Standard				
¼, ⅜ in.	62	38½	....	....
½ in.	67	47½	....	....
¾ to 1½ in.	72	52¼	....	....
2 in.	72	52½	68	48½
2½ to 4 in.	72	52½	71	51½
4½, 5, 6 in.	....	....	69	49½
7, 8, 10 in.	....	....	66	44½
X Strong P. E.				
¼, ⅜ in	55	38½	....	....
½ in	62	45½	....	....
¾ to 1½ in.	66	49½	....	....
2, 2½, 3 in.	67	50½	....	....
2 in.	....	....	62	45½
2½ to 4 in	....	....	65	48½
4½, 5, 6 in.	....	....	65	48½
7, 8 in	....	....	58	39½
XX Strong P. E.				
½ to 2 in	43	26½	....	....
2½ to 6 in.	....	....	42	25½
7 to 8 in.	....	....	39	20½
Genuine Wrot Iron.				
⅜ in	56	32½	....	....
½ in	61	41½	....	....
¾ to 1½ in.	66	46½	....	....
2 in.	66	46½	62	42½
2½, 3 in.	66	46½	65	45½
3½, 4 in.	....	....	65	45½
4½, 5, 6 in.	....	....	62	42½
7, 8 in.	....	....	59	37½

Wrought Nipples.

4 in. and under	77½%
4½ in. and larger	72%
4 in. and under, running thread.	57½%

Standard Couplings.

4 in. and under	60%
4½ in. and larger	40%

MILLED PRODUCTS.

Sq. & Hex Head Cap Screws	65 & 5%
Sq. Head Set Screws	70 & 5%
Rd. & Fil. Head Cap Screws	45%
Flat & But. Head Cap Screws	40%
Finished Nuts up to 1 in.	70%
Finished Nuts over 1 in.	70%
Semi-Fin. Nuts up to 1 in.	70%
Semi-Fin. Nuts over 1 in.	72%
Studs	65%

METALS.

	Montreal.	Toronto.
Lake Copper, carload	$21 50	$20 75
Electrolytic copper	21 25	20 50
Castings, copper	21 00	20 50
Tin	45 00	43 00
Spelter	21 00	18 00
Lead	6 75	7 00
Antimony	42 00	40 00
Aluminum	70 00	65 00

Prices per 100 lbs.

BILLETS.

	Per Gross Ton
Bessemer billets, Pittsburgh	$30 00
Open-hearth billets, Pittsburgh	31 00
Forging billets, Pittsburgh	52 00
Wire rods, Pittsburgh	40 00

NAILS AND SPIKES.

Standard steel wire nails, base	$2 80	$2 85
Cut nails	2 90	2 90
Miscellaneous wire nails	75 per cent.	
Pressed spikes, ⅝ diam., 100 lbs.	3 25	

BOLTS, NUTS AND SCREWS.

	Per Cent.
Coach and lag screws	60 and 5
Stove bolts	82½
Plate washers	40
Machine bolts, ⅜ and less	65
Machine bolts, 7-16 and over	50
Blank bolts	50-7½
Bolt ends	50-7½
Machine screws, iron, brass	35
Nuts, square, all sizes	3¾c per lb off
Nuts, hexagon, all sizes	4¼c per lb. off
Iron rivets	67½
Boiler rivets, base, ¾-in. and larger	$3.75
Structural rivets, as above	3.75
Wood screws, flathead, bright	85, 10, 10 p.c. off
Wood screws, flathead, brass	67½ p.c. off
Wood screws, flathead, bronze	60 p.c. off

LIST PRICES OF W. I. PIPE.

Standard. Nom. Diam.	Standard. Price. per ft.	Extra Strong. Sizes Ins.	Extra Strong. Price per ft.	D. Ex. Strong. Size Ins.	D. Ex. Strong. Price per ft.
⅛in	$.05½	⅛in	$ .12	½	$ .32
¼in	.06	¼in	.07½	¾	.35
⅜in	.06	⅜in	.07½	1	.37
½in	.08½	½in	.11	1¼	.52½
¾in	.11½	¾in	.15	1½	.65
1 in	.17½	1 in	.22	2	.91
1¼in	.23½	1½in	.30	2½	1.37
1½in	.27½	1½in	.36½	3	1.86
2 in	.37	2 in	.50½	3½	2.30
2½in	.58½	2½in	.77	4	2.76
3 in	.76½	3 in	1.03	4½	3.26
3½in	.92	3½in	1.25	5	3.86
4 in	1.09	4 in	1.50	6	5.32
4½in	1.27	4½in	1.80	7	6.35
5 in	1.48	5 in	2.08	8	7.25
6 in	1.92	6 in	2.86	....	....
7 in	2.38	7 in	3.81	....	....
8 in	2.50	8 in	4.34	....	....
8 in	2.88	9 in	4.90	....	....
9 in	3.45	10 in	5.49	....	....
10 in.	3.20	....	....	....	....
10 in.	3.50	....	....	....	....
10 in.	4.12	....	....	....	....

COKE AND COAL

Solvay Foundry Coke $6.25	
Connellsville Foundry Coke 5.65	
Yough Steam Lump Coal 3.63	
Penn. Steam Lump Coal 3.63	
Best Slack 2.99	

Net ton f.o.b. Toronto.

COLD DRAWN STEEL SHAFTING.

At mill 25%	
At warehouse 20%	

Discounts off new list. Warehouse price at Montreal and Toronto.

MISCELLANEOUS

Solder, half-and-half 0.23½	
Putty, 100-lb. drums 2.70	
Red dry lead, 100-lb. kegs, per cwt. 9.65	
Glue, French medal, per lb. 0.15	
Tarred slaters' paper, per roll ... 0.95	
Motor gasoline, single bbls., gal. ..0.25½	
Benzine, single bbls., per gal. ... 0.25	
Pure turpentine, single bbls. 0.87	
Linseed oil, raw, single bbls. 0.87	
Linseed oil, boiled, single bbls.... 0.90	
Plaster of Paris, per bbl. 2.50	
Plumbers' Oakum, per 100 lbs... 4.50	
Lead Wool. per lb. 0.11	
Pure Manila rope 0.16	
Transmission rope, Manila 0.20	
Drilling cables, Manila 0.17	
Lard oil, per gal 0.73	
Union thread cutting oil 0.60	
Imperial quenching oil.......... 0.35	

POLISHING DRILL ROD

Discount off list, Montreal and Toronto 40%

PROOF COIL CHAIN.

¼ in. $9.00	
5-16 in. 5.90	
⅜ in. 4.95	
7-16 in. 4.55	
½ in. 4.00	
9-16 in. 4.20	
⅝ in. 4.10	
¾ in. 3.95	
⅞ in. 3.80	
1 inch 3.70	

Above quotations are per 100 lbs.

TWIST DRILLS.

	%
Carbon up to 1½ in.	55
Carbon over 1½ in.	25
High Speed	
Blacksmith	55
Bit Stock ....60 and 5	
Centre drill	20
Ratchet	20
Combined drill and c.t.s.k.	15

Discounts off standard list.

REAMERS

	%
Hand	25
Shell	25
Bit Stock	25
Bridge	65
Taper Pin	25
Centre	25
Pipe Reamers......	80

Discounts off standard list.

IRON PIPE FITTINGS.

Canadian malleable, A, 25 per cent.; B and C, 35 per cent.; cast iron, 60; standard bushings, 60 per cent.; headers, 60; flanged unions, 60; malleable bushings, 60; nipples, 75; malleable, lipped unions, 65.

TAPES

Chesterman Metallic, 50 ft. $2.00	
Lufkin Metallic, 603, 50 ft. 2.00	
Admiral Steel Tape, 50 ft. 2.75	
Admiral Steel Tape, 100 ft. 4.45	
Major Jun., Steel Tape, 50 ft. ... 3.50	
Rival Steel Tape, 50 ft. 2.75	
Rival Steel Tape, 100 ft. 4.45	
Reliable Jun., Steel Tape, 50 ft. .. 3.50	

SHEETS.

	Montreal	Toronto
Sheets, black, No. 28	$3 50	$3 50
Canada plates, dull.		
52 sheets	3 25	3 25
Canada Plates, all bright..	4 60	4 75
Apollo brand, 10¾ oz.		
galvanized	5 50	5 50
Queen's Head. 28 B.W.G.	6 00	6 00
Fleur-de-Lis, 28 B. W. G...	5 75	5 75
Gorbal's Best, No. 28 ...	6 10	6 10
Viking metal, No. 28 ...	5 25	5 25
Colborne Crown, No. 28..	5 70	5 80
Premier No. 28	5 40	5 50
Premier, 10¾ oz.	....	5 75

BOILER TUBES.

Size	Seamless	Lapwelded
1 in.	$14 25	.....
1¼ in.	15 00	.....
1½ in.	15 00	.....
1¾ in.	15 00	.....
2 in.	15 00	10 00
2¼ in.	16 50	11 00
2½ in.	17 50	12 85
3 in.	25 00	13 20
3½ in.	28 00	16 25
4 in.	33 00	20 75

Prices per 100 feet, Montreal and Toronto.

WASTE.

WHITE.	Cents per lb.
XXX Extra	0 11½
X Grand	0 11
XLCR	0 10¼
X Empire	0 09½
X Press	0 08¾
COLORED.	
Lion	0 07¾
Standard	0 07
Popular	0 06¼
Keen	0 05½
WOOL PACKING.	
Arrow	0 17
Axle	0 12
Anvil	0 09
Anchor	0 07
WASHED WIPERS.	
Select White	0 08½
Mixed Colored	0 06¼
Dark Colored	0 05¼

This list subject to trade discount for quantity.

BELTING RUBBER

Standard	.50%
Best grades	.30%

BELTING—NO. 1 OAK TANNED.

Extra heavy, single and d'ble, 40 & 10%	
Standard 50%	
Cut leather lacing, No. 1........$1.20	
Leather in sides 1.10	

ELECTRIC WELD COIL CHAIN B.B.

⅛ in. $12.75	
3-16 in. 8.85	
¼ in. 6.15	
5-16 in. 4.90	
⅜ in. 4.05	
7-16 in. 3.85	
½ in. 3.75	
⅝ in. 3.60	
¾ in. 3.60	

Prices per 100 lbs.

PLATING CHEMICALS

Acid, boracic .... $ .15	
Acid, hydrochloric .... .05	
Acid, hydrofluoric .... .06	
Acid, nitric .... .10	
Acid, sulphuric .... .05	
Ammonia, aqua .... .08	
Ammonium carbonate .... .15	
Ammonium chloride .... .11	
Ammonium hydrosulphuret ..... .35	
Ammonium sulphate .... .07	
Arsenic, white .... .10	
Copper sulphate .... .10	
Cobalt sulphate .... .50	
Iron perchloride .... .20	
Lead acetate .... .16	
Nickel ammonium sulphate ..... .10	
Nickel carbonate .... .50	
Nickel sulphate .... .15	
Potassium carbonate .... .40	
Potassium sulphide (substitute).. .20	
Silver chloride ....(per oz.) .65	
Silver nitrate ....(per oz.) .45	
Sodium bisulphite .... .10	
Sodium carbonate crystals ...... .04	
Sodium cyanide, 127-130% ..., .35	
Sodium hydrate .... .04	
Sodium hyposulphite (per 100 lbs.) 3.00	
Sodium phosphate .... .14	
Tin chloride .... .45	
Zinc chloride .... .20	
Zinc sulphate .... .07	

Prices Per Lb. Unless Otherwise Stated.

ANODES

Nickel .... .47 to .52	
Cobalt .... 1.75 to 2.00	
Copper .... .22 to .25	
Tin .... .45 to .50	
Silver .... .55 to .60	
Zinc .... .22 to .25	

Prices Per Lb.

PLATING SUPPLIES

Polishing wheels, felt .... 1.50 to 1.75	
Polishing wheels, bullneck. .80	
Emery in kegs .... 4½ to .06	
Pumice, ground .... .05	
Emery glue .... .15 to .20	
Tripoli composition .... .04 to .06	
Crocus composition ...... .04 to .06	
Emery composition ...... .05 to .07	
Rouge, silver .... .25 to .50	
Rouge, nickel and brass... .15 to .25	

Prices Per Lb.

The General Market Conditions and Tendencies

This section sets forth the views and observations of men qualified to judge the outlook and with whom we are in close touch through provincial correspondents

Montreal, Que., Dec. 13, 1915.—Reports from all sources continue to show that satisfactory conditions prevail throughout the steel, iron and metal industries, and while munitions are the chief feature of the activity, many other lines are in more or less good demand.

Pig Iron

Local conditions in the pig iron market are little changed from the previous week. Production continues at maximum capacity to supply the demands made upon the steel makers. In spite of the high prices, considerable inquiries are coming in from foundries. The question of transportation of both ore and iron is having a tendency to advance prices on some grades. A disturbing factor for the coming year is the possibility of a shortage in vessel tonnage for the shipment of ore.

Steels

The tension in the steel situation shows little indication of relaxation; in fact, the pressure upon producers seems to be increasing, and the period of future deliveries is constantly being extended. While bars and billets are principally in demand, the inquiries for bars, plates, etc., continue to increase.

Many mills are limiting orders to a proportion of customers' requirements, and in some instances inquiries are not even considered. Producers of steel have business booked for many months ahead, and in a great number of cases orders for second and third quarters of 1916 cannot be taken on.

The continued scarcity of tungsten keeps the figures for high-speed steel excessively high, and little relief is immediately expected. However, by the spring of next year, the situation may be somewhat relieved. Present quotations are $2.50 to $3.25 per pound.

Metals

The general condition of the metal market is dull, with prices firm. Local spelter shows a slight advance, while aluminum records another jump of 5c per pound. Present conditions will likely prevail until the close of the year.

Copper.—The lively condition of the copper market for the past few weeks has now ceased, and indications are that little change will be noted for the remainder of the month. Slight declines are quoted in the United States markets, but local prices are holding firm.

Tin.—No new developments have taken place during the past week, and the general dullness continues, with the possibility of a further slight decline before the opening of the New Year. A considerable supply of tin is in sight, and this has weakened the foreign markets. However, the local dealers are quoting last week's figures of 45 cents per lb.

Spelter.—Conflicting reports are being circulated regarding the spelter situation, and it would seem that speculators are trying to force prices up. It is generally conceded that recent advances are due to this, more so than supply and demand. However, the quiet condition at present tends to retain firmness, and, while foreign prices have declined somewhat, local dealers have advanced their quotations 1c per pound, which leaves the present price at 21c.

Lead.—The outlook from present conditions are that the market will show

CANADIAN GOVERNMENT PURCHASING COMMISSION

The following gentlemen constitute the Commission appointed to make all purchases under the Dominion $100,000,000 war appropriation:—George F. Galt, Winnipeg; Hormidas Laporte, Montreal; A. E. Kemp, Toronto. Thomas Hilliard is secretary, and the commission headquarters are at Ottawa.

weakness in a short time, a decline being generally expected. Last week's quotation of 6¾c holds firm for the present.

Antimony.—The market is steady and prices are firm at 42 cents per lb.

Aluminum.—The demand for aluminum continues to boost the price, and inquiries are numerous. Local dealers have advanced their quotations from 65 cents to 70 cents per pound.

Old Material.—Local scrap conditions remain unchanged, with a fair amount of transactions passing. Heavy melting steel scrap is quite active, and the demand is increasing. Present prices show an advance of $10 per ton over last week. The shortage of aluminum has advanced the price of its scrap to 27 cents per pound.

Toronto, Ont., Dec. 14.—There is little change to note in the general situation. The revival in trade continues and the outlook is distinctly favorable. The export trade is increasing in volume while domestic trade has improved considerably of late. Railways report substantial increases in earnings, and bank clearings exceed last year as also those of the previous year, which are favorable indications as to the increase in the volume of trade, and improvement in conditions generally. The shortage of tonnage and high ocean freight rates are a serious inconvenience to importers and exporters, and have not only increased the cost of certain products but have affected deliveries also.

The steel market continues very strong with every possibility of higher prices. The demand for steel for shells is increasing and is greater than the Canadian mills can produce. Preparations are being made for turning out the large-calibre shells which will call for a considerable increase in tonnage. The machine tool trade is active and dealers report very satisfactory business. In the metal market, tin and spelter have declined, otherwise prices are unchanged. The market generally is quiet.

Steel Market

The extraordinary activity in the steel trade continues and there is no sign of any recession. The steel companies have more business than they can handle and are booked up for months ahead. The demand is so much greater than the Canadian mills can supply that concerns working on shell contracts are obliged to buy raw materials from the States. It is stated that several large contracts for steel forgings and round steel bars which have been under negotiation between Canadian concerns and American mills are held in abeyance temporarily. It is understood, however, that the contracts will eventually be placed in the States. Under prevailing conditions, prices of steel are bound to advance, the market being very strong. Boiler plates have advanced again, and higher prices for domestic steel and iron bars are expected any time. There is no change in the situation with regard to American bars, shapes, etc., and prices are still withdrawn.

The galvanized sheet market is very irregular. Many makers have discontinued quoting on sheets on account of the fluctuations in the spelter market. Prices of black sheets continue to advance and all other materials used in the manufacture of galvanized sheets are also higher. Prices of galvanized sheets are very fair, and are expected to go higher.

Prices continue to advance in the States and the export demand is increasing. The French government is enquiring for thousands of tons of round steel bars. Many of the mills are increasing their capacity, but are as far behind as ever on delivery. Steel bars are quoted at 1.70c base, and iron bars 1.80c base, Pittsburgh. Billets continue to advance on heavy demand. Bessemer billets are now quoted at $30 and open-hearth billets at $31 Pittsburgh. Forging billets are unchanged at $52, and wire rods are higher at $40 Pittsburgh.

Pig Iron

The pig iron market is very strong and prices continue to advance. A large percentage of the tonnage is going into steel, there being little demand for foundry grads. "Hamilton" and "Victoria" brands have advanced $1, and are now quoted locally at $24 per ton. The market for pig iron in the States is becoming excited and prices are advancing. "Grey Forge" is now quoted at $17.95 Pittsburgh. It is understood that quotations on "Lake Superior" iron ore for the coming season have been made. "Mesaba Bessemer" is quoted at $4.25, which is 80c above this year's basis. On "Mesaba non-Bessemer," 70c advance over this year is asked.

Old Materials

There is little change to note with regard to the situation in old materials. The market generally is firm with higher prices for heavy melting steel, aluminum and lead, while the various grades of copper have also advanced slightly. The market is not very active but there is a fair demand for copper and heavy melting steel. There is a continued scarcity of aluminum; this has strengthened the market for scrap metal.

Machine Tools

The market is very active and there is a good demand for machinery for shells, principally for the larger sizes. Some orders for 8-in. shells have been placed, and the concerns thus favored are getting their plants in shape for machining them. There appears to be a possibility of the situation as regards getting machine tools, being considerably relieved. There is a growing belief in the American market that purchases of machine tools for shipment to England will in the future be of comparatively small proportions. It is believed that munition making plants of Great Britain are now capable of meeting all requirements; the demand for tools is, therefore, less urgent. In this event, machine tools will be easier to obtain for this market and deliveries will be better. It is understood that Russia is buying a large amount of equipment, not only lathes, but practically a general line of tools.

Supplies

A few price changes have to be noted this week. Solder is down ½c, due to the continued weakness in the tin market. Linseed oil market is very strong and prices have advanced 2c, the new quotation being 87c for raw and 90c for boiled, per gallon. Turpentine is 2c high-

er and is now quoted at 87c per gallon. High-speed twist drills keep on advancing and the situation is such that it is impossible to give a firm quotation. Business continues very good, there being an active demand for supplies from shell plants.

Metals

The market is quiet this week and lower levels for tin and spelter have again to be noted, other metals being unchanged. Copper is quiet, although the market is in a strong position. The lead market is quiet but firm. There is no change in the antimony situation, and the aluminum market is featureless. There is a continued active demand for

metals for munitions, but ordinarily business is quiet.

Tin.—Although the primary markets are firm, tin is weaker locally and has declined 2c. There is a fair demand, but stocks are accumulating, which may depress the market. Tin is quoted locally at 43c per pound.

Copper.—The market is in a strong position, although quiet and unchanged. There is apparently a lull in demand, both for domestic and export. "Lake Copper" is quoted at 20¾c per pound.

Spelter.—The market is in a very peculiar condition, and it is difficult to size up the situation. There is good inquiry but sellers are not showing much disposition to do business. Spelter has declined 2c locally and quotations are more or less nominal at 18c per pound. Zinc ore is quoted from $80 to $105, Joplin, Mo.

Lead.—The market is quiet but firm at the "Trust" equivalent of 5.25c New York. The ore is quoted at $70 to $73, Joplin, Mo. Lead is unchanged locally at 7c per pound.

Antimony.—Conditions in the antimony market are unchanged. Spot stocks are scarce but futures are being freely offered at a concession of about 2c less than the spot market. Local quotations are unchanged at 40c per pound.

Aluminum.—There is no change in the situation. Supplies are scarce and the demand continues heavy. Quotations are nominal at 65c per pound.

CANADIAN EXPORTS TO GREAT BRITAIN

AN increase in imports from Canada of $34,697,000 and a decrease in exports to Canada of $20,498,000 during the first nine months of the present fiscal year is shown by figures for that period received by the Trade and Commerce Department from Commissioner Ray, of Birmingham. The total imports from Canada were $138,917,000, and exports to the Dominion $49,407,000.

Details are also furnished of the recent steps taken by the British authorities in respect to the requisitioning of all ships of British register, which regulations closely affect Canada. These show that rumors that the Government contemplates the requisitioning of the entire British mercantile marine are without foundation, but that power has been taken to deal, by the requisitioning of a sufficient number of vessels, with cases where an emergency of national importance exists at any particular market owing to the absence of tonnage, and, further, to regulate the employment of British shipping in the carriage of cargoes between foreign parts by means of licenses.

COMPENSATION ACT ADJUSTMENTS

A NUMBER of important changes in the grouping of industries under the Workmen's Compensation Act have been decided upon by the Workmen's Compensation Commission. The chief changes are a regrouping of the iron industries, in response to a request from employers; the separation of the manufacture of explosives from all other industries, and the uniting of all the building trades in one class. There has also been a rearrangement of railway, canal, roadmaking and bridgebuilding industries.

Theatres and moving picture houses, and the operation of elevators not in industries under Schedule 1, which had formerly been specifically included, are now excluded. The Board has also made the Act apply expressly to a number of industries hitherto considered to be covered by the general scope of the Act, but not expressly mentioned. The exclusion of machine shops, cabinet shops, and tinsmith shops with less than a specified number of employees has also been removed, and exclusion from the schedule is permitted in regard to certain wood operations carried on for the most part by farmers and settlers, and to repairing or small building operations carried on by owners of rented houses or buildings. The changes come into effect on January 1, and will not affect employers during the present year.

The Board announces that employers must send in their next pay roll statement by the 20th of January. In the meantime a circular explaining the changes in detail will be prepared for distribution, together with the pay roll form for the coming year.

DOMINION COAL CO. ST. LAW-RENCE SEASON

THE quantity of coal transported to St. Lawrence ports by the Dominion Coal Co. during the 1915 season of navigation was about 1,600,000 gross tons. In addition to this, something over 100,000 tons will have to be handled through Portland over the ensuing winter months, making a total for St. Lawrence ports of 1,700,000 tons.

A slightly larger quantity was transported to Montreal during the 1914 season, but the difference is accounted for by the fact that steamers formerly taking large bunker supplies to St. Lawrence ports were diverted to Sydney in the season just past, and bunkered either on their inward or outward voyage.

Steamers employed in the coal-carrying trade made 310 trips to St. Lawrence ports during the season of navigation just ended, and vessels from the Upper Lakes, as well as the seven seas, were secured in order to carry this large amount of coal at a time notable for shortage of tonnage.

POWER DEVELOPMENT AT MUSKOKA

THE Hydro-Electric Power Commission of Ontario has under way the extension of a power plant on the south branch of the Muskoka River. This plant, located at South Falls, and formerly operated to supply the town of Gravenhurst, has a single unit with a capacity of 450 k.w. It is being enlarged by the addition of a 750 k.v.a. unit. Excavation work was recently commenced and most of the equipment has been ordered.

Provision was made in the design of the existing plant for such an extension, and the installation of the latter does not alter the previous arrangement.

A wood stave pipe line, 100 feet long, will connect the existing headworks with about 60 feet of steel penstock leading to the new unit. The headgate mechanism, steel penstock, turbine, governor and relief valves are all being supplied by the William Hamilton Co., of Peter-

borough, Ont., the generator and the transformers by the Canadian General Electric Co., Ltd., Peterborough, and the wood pipe line by the Pacific Coast Wood Stave Pipe Co. The latter is being laid by the commission. Orders for the switchboard equipment have not yet been placed.

The extended plant will supply the towns of Gravenhurst, Bracebridge and Huntsville.

NEW ELECTRIC STEEL FURNACE PLANT

THE Canadian Electro-Products Co., whose incorporation with a capital of $500,000 was announced recently, expects to have a new electric furnace plant for the manufacture of high-grade steel in operation in Montreal in about a month

or six weeks. The plant is being designed on the basis of two units, each of a capacity of 25 tons of steel a day, and at least one of these units should be in operation within the time mentioned.

The authorized capital of $500,000, it is understood, will be half in preferred and half in common shares, but only a portion of the authorized amount will be necessary for the initial installation. Probably $100,000 will be required to equip the plant to start with, and this is being subscribed privately.

J. S. Norris, of the Montreal Light, Heat & Power Co. and Messrs. Howard Murray and Julian C. Smith, of the Shawinigan Water and Power Co., are among those behind the new venture and will be members of the board of directors when organization is completed. The company, however, is a private enter-

prise, and will not be a subsidiary of either Montreal Power or Shawinigan Power. However, as the operation of the new plant, when both units are working, will call for about 4,000 h.p., naturally both the older companies will benefit, one as the distributor and the other as the producer of the power required.

The new venture represents a further extension of the electric furnace idea, which is playing so large a part in the increased consumption of power in this province. It has been applied to the aluminum and carbide industries in the Shawinigan district on a large scale. As applied to steel on the lines proposed, it is a new proposition here, but similar plants are in operation at Welland and Sherbrooke.

DOMINION TRADE RETURNS

AN expanding revenue and a decreased expenditure on consolidated account are shown in the November financial statement of the Dominion issued on Dec. 10. The total revenue to the end of November was $104,756,305.25, as against $90,468,002.68 for the corresponding period of last year, increases appearing under all heads, with the exception of excise. The consolidated fund expenditure dropped from $75,708,627 to $65,345,503.

Expenditure on capital account outside of the war outlay also shows a substantial reduction. The 1914 column shows no war expenditure, but the outlay last month under this head is given as $13,155,797, bringing the capital expenditure on war for the fiscal period up to $66,514,955. The total capital expenditure to the end of November was $91,475,889, as against $28,231,933 for the same period last year. The gross debt increased by $202,225,076, and stood at $829,377,292 at the close of the month. The funded debt payable in Canada rose from $774,060 in November, 1914, to $8,725,450 in November, 1915. The funded debt payable in England increased from $329,020,000 to $342,703,302. Temporary loans rose from $20,573,333 to $165,007,017.

Increases in the sinking funds and miscellaneous and banking accounts brought the total assets of the Dominion up to $327,709,125, as against $262,308,968, the total net debt thus standing at $501,668,167, as compared with $364,843,247. The monthly increase in the net debt was $9,139,675, as against $12,167,848, a decrease of $3,028,175.

Stratford, Ont.—The directors of the People's Telephone Co. have decided to purchase the Forest plant of the Bell Telephone Co. It is expected that the change will be made at the commencement of the New Year.

ONTARIO MINERAL OUTPUT

RETURNS made to the Ontario Bureau of Mines for the nine months, ending September 30, 1915, show an increase in value of gold of $1,884,093, and a decrease in value of silves of $2,051,760. It is pointed out that the increase in the production of gold amounts to one-third.

The gold districts of Northern Ontario, the report says, are fulfilling the prediction made several years ago that they would make good the loss caused by the waning of the silver mines of Cobalt. Thus, the combined value of the gold and silver output of the first nine months of the present year was only $167.661 less than for the same portion of 1914, notwithstanding the fact that the yield of silver fell off over 20 per cent. Part of this decrease is due to the low prices which prevailed during the whole nine months, but which made a sharp recovery in November.

The Sudbury mines are being worked to the maximum capacity, and the production of nickel for the nine months nearly equals the largest previous output for a full year. Over 75 per cent. of the output is made by the Canadian Copper Co., but the operations of the Mond Co. are now more extensive than formerly, and its output has correspondingly increased.

The yield of copper was also much greater than in the corresponding period of 1914 and nearly equalled the total output of that year.

New Incorporations

The Montreal Steel & Foundry Co. has been incorporated at Ottawa, with a capital of $150,000, to manufacture all kinds of machinery and mechanical specialties, appliances and instruments, at Montreal, Que. Incorporators: Emilien Gadbois, Joseph Marechal Nantel, and Charles G. Derome, all of Montreal, Que.

The Barrymore Cloth Co. has been incorporated at Toronto, with a capital of $250,000, to take over as a going concern the cloth manufacturing portion of the business of the Toronto Manufacturing Co., of Toronto, Ont. Incorporators: Edmund Percival Brown and Wm John McWhinney, of Toronto.

The Wood Products Co. has been incorporated at Toronto, with a capital of $100,000, to engage in the destructive distillation of wood and to manufacture charcoal, wood alcohol and all other products. Head office at Toronto. Incorporators: Arthur Wellesley Holmestead and James Leith Ross, of Toronto.

The International Steel Corporation, Ltd., has been incorporated at Ottawa, with a capital of $100,000, to manufacture, produce and deal in iron, steel and all other metals, at Toronto. Incorporators: James Richardson Roaf, Wm. Graham and John F. Mordon, all of Toronto, Ont.

INDUSTRIAL ᴬᴺᴰ CONSTRUCTION NEWS

Establishment or Enlargement of Factories, Mills, Power Plants, Etc.; Construction of Railways, Bridges, Etc.; Municipal Undertakings; Mining News.

Engineering

Glace Bay, N.S.—The town council propose installing a pumping plant.

Hamilton, Ont.—The Monarch Machine Tool Co. will start work shortly on a war order.

Hamilton, Ont.—The Steel Co. of Canada, Ltd., will make considerable extensions to their plant.

Georgetown, Ont.—The Glass Garden Builders, Ltd., will erect a factory to cost about $25,000.

Montreal, Que.—The Williams Mfg. Co. will build a foundry at their plant on St. James street.

Hamilton, Ont.—The Dominion Steel Castings Co. will build an extension to their plant.

Bedford, Que.—A pumping plant will be installed in connection with the waterworks extensions now in progress.

Toronto, Ont.—The Board of Control will call for tenders for the supply and installation of a boiler at the municipal abattoir.

London, Ont.—The Battle Creek Toasted Corn Flake Co., Ont., will install electrical equipment to operate its machinery.

Halifax, N.S.—Work has been started on the erection of a steel plant for the Williston Steel & Foundry Co. Approximate cost, $16,000.

New Glasgow, N.S.—The Nova Scotia Steel & Coal Co. are building an addition to their steel plant at Sidney Mines, estimated to cost $100,000.

Fort Frances, Ont.—Russel Bros., machinists, have secured a contract for shells and will remove to Port Arthur where a plant will be equipped with the necessary machinery.

Edmonton, Alta.—G. W. Farrell & Co., of Montreal, Que., has signed a contract to supply power to the city. It will build a power plant and an electric railway to cost $6,000,000.

Arnprior, Ont.—The Galetta Power Co. are considering developing a water power at Long Rapids on the Madawaska river. A minimum of 13,000 h.p. could be developed at that point, but it would be difficult at present to dispose of so much power.

Quebec, Que.—Fire damaged the Eastern Canada Steed Co. plant, St. Malo, recently. The cause of the fire is not exactly known, but it is supposed that it originated through an explosion in the furnace room. The damage, which is said to be around $100,000, is covered by insurance.

Grimsby, Ont.—T. D. G. Bell and his associates have formed a new company to be known as Bells, Limited, for the purpose of making war munitions. The new company has rented the necessary space and plant from the Bell Fruit Farms, Ltd., and the new machinery to make munitions will be installed in a portion of the factory.

Montreal, Que.—The manufacture of steel billets will shortly be begun by Montreal Power-Shawinigan interests who have purchased the Record Foundry Building, and are now installing the necessary equipment. The concern, known as the Canadian Electro-Products Co., has secured its charter and machinery, including an electric furnace.

Stratford, Ont.—At a meeting of the Finance Committee of the City Council on Dec. 5, a committee composed of Chairman Ald. Henry and Aldermen Forbes, Mantle and Pauli, was appointed to sell the machinery which is being cleaned out of the basement in the Kemp factory, to make room for the recruits to the 110th Batt.

Municipal

Collingwood, Ont.—The proposed by-law in connection with Wilson Bros. has been withdrawn.

Toronto, Ont.—The Board of Control has decided to submit the hydro-radial to the ratepayers on Jan. 1.

Windsor, Ont.—A by-law is being prepared to sanction the expropriation of the Windsor Gas Co.

Islington, Ont.—Etobicoke Township ratepayers will vote New Year's Day on the proposed system of Hydro-radials.

Petrolia, Ont.—A by-law in connection with the proposed sugar refinery will be submitted to the ratepayers on Jan. 3.

Berlin, Ont.—The city council have decided to submit the hydro-radial by-law to the electors at the January elections.

Chatham, Ont.—A by-law will be voted on by the ratepayers to sanction an expenditure of $10,000 on fire-fighting apparatus.

Chatham, Ont.—A by-law will be submitted to the ratepayers relative to granting concessions to the Dominion Sugar Co.

Baden, Ont.—The township council have decided to submit the hydro-radial by-law to the ratepayers of Wilmot Township on Jan. 3.

Chatham, Ont.—A by-law will be submitted to the ratepayers at the January elections regarding concessions to the Gray-Dart Automobile Co.

Sarnia, Ont.—A by-law will be voted on by the ratepayers on Jan. 3 to raise $120,000 for the purchase of the Sarnia Gas and Electric Light Co. plant.

Sarnia, Ont.—A by-law will be submitted to the ratepayers on Dec. 29 to authorize raising $12,000 to pay for waterworks extensions.

Tara, Ont.—A by-law will be submitted to the ratepayers on Jan. 3 to authorize an expenditure of $7,500 on a hydro-electric distribution plant.

Sherbrooke, Que.—The City Council contemplates alterations and improvements to the water power, electric transmission and lighting plants.

Listowel, Ont.—A by-law will be voted on by the ratepayers on Jan. 3, to authorize a loan of $12,000 to a company who propose to establish a boot factory here.

Cobden, Ont.—The town council will submit a by-law on December 27 to provide for the construction of a plant for the generation and distribution of power. Estimated cost, $20,000.

Princeville, Que.—The town council contemplates the erection of a pump station and the purchase of a 4-inch and 6-inch cast iron pipe, valves, pumps and hydrants.

Swift Current, Sask.—The City Council will engage an engineer to make a thorough examination of the pumping and waterworks system and advise as to what will be required to make the system thoroughly efficient.

Lindsay, Ont.—A by-law will be voted on at the January elections to authorize the granting of a loan of $15,000 to a

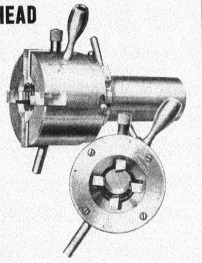
If what you want is not advertised in this issue consult the Buyers' Directory at the back.

company who propose establishing a plant to manufacture wood alcohol, charcoal and acetic acid. Thomas Hodgson is promoting the company.

Toronto, Ont.—Parks Commissioner Chambers has recommended to Board of Control that three stand-pipes, the necessary quantity of fire hose, and three portable extinguishers be installed at the Process Building at the Exhibition grounds, where the soldiers are accommodated. The cost of the equipment will be $255.

Petrolea, Ont.—The council on Dec. 7 gave first and second readings to the by-law for $21,500 to assist the Western Sugar Refining Co., which will locate here, if this inducement is granted. A by-law to raise $25,000 to purchase stock in the company was twice read. These issues will be voted upon by the ratepayers January 3.

Electrical

St. Thomas, Ont.—The Michigan Central Railway will electrify their terminal at this place.

Hull, Que.—The city council are considering extensions to the lighting system, estimated to cost $20,000.

Biddulph Township, Ont.—A by-law has been passed providing for the installation of a hydro-electric system at an approximate cost of $3,500.

Burgessville, Ont.—A by-law to provide for the construction of a hydro-electric system at an approximate cost of $3,500 will shortly be submitted by the township council of South Norwich.

General Industrial

St. Catharines, Ont.—Fire on Dec. 7 destroyed the Maple Leaf Milling Co.'s flour mill here. The loss is estimated at $75,000, which is fully covered by insurance.

Davidson, Sask.—The Canadian Elevator Co. will build an extension to their elevator. The North Star Elevator Co. and the British American Elevator Co. will also enlarge their mills.

Vancouver, B.C.—The Canada Potash and Algin Co. will shortly work on a new plant for making iodine and gum algin. These materials will be made from kelp, which is found in large quantities along the coast.

Sackville, N.B.—The A. E. Wry Standard, Ltd., factory was damaged by fire on December 5, the loss being estimated at $35,000, of which $20,000 is covered by insurance. The company manufacture harness and other leather goods.

Toronto, Ont.——The factory of T. Crowley & Co., manufacturers of picture frames, on McDonnell Square, was badly damaged by fire on Dec. 12. The damage is placed at about $1,500 to the building and $7,000 to the stock and contents. The loss is covered by insurance.

Contracts Awarded

The Otis-Fensom Co., Hamilton, Ont., have been awarded a contract for an elevator by the Patterson Mfg. Co., of Montreal, Que.

Port Moody, B.C.—The contract for the construction of a waterworks system has been awarded to the Robertson-Godson Co., Vancouver, B.C., at $37,000.

Toronto, Ont.—The Board of Control has awarded a contract for the supply and installation of valves, steam piping, special steel castings and lagging at main pumping station to Purdy, Mansell, Ltd., at $19,995.

Brockville, Ont.—The Public Utilities Commission at a meeting on Dec. 7 awarded the contract for the new filtration plant to the Roberts Filtration Company, of Darby, Pa., for $86,000, and also requested the Town Council to provide $115,000 to cover the filtration plant, low lift pumps and intake cost.

Tenders

St. Hyacinthe, Que.—Tenders will be received up to January 11, for a mechanical filter plant. Plans and specifications may be obtained at the office of Hector Cadieux, city engineer.

Halifax, N. S.—Tenders will be received by the Governor of the Province of Macao, up till January 8, 1916. for the supply of a steel, self-propelling dredge for the use of the Macao Harbor Works. Full particulars may be obtained at the office of Fred H. Oxley, Consul for Portugal, Keith Bldg., Halifax, N.S.

Winnipeg, Man.—Tenders will be received up to Monday, December 27, 1915, for the supply of one saddle tank locomotive to the Greater Winnipeg Water Commission. Specifications and form of tender may be inspected at the offices of the district, 901 Boyd Building, Winnipeg, Man.

Winnipeg, Man.—Tenders addressed to the undersigned will be received up to Wednesday, December 29, 1915, for the supply of twenty twenty-yard automatic air-dump cars. Specifications and form of tender may be obtained and form of contract may be inspected at the office of the district. S. H. Reynolds, Chairman of the Greater Winnipeg Water Commissioners, 901 Boyd Building, Winnipeg, Man.

Trade Gossip

The Shawinigan Electro Metals Co. have increased the capital stock of the concern to $200,000.

The International Steel Corporation, Ltd., have changed the name of the concern to Canadian Iron Ores, Ltd.

The Turnbull Elevator Co., Toronto, have been awarded a contract for the elevators for the Crompton Corset Co.'s new warehouse.

Simcoe, Ont.—The Town Council will submit a by-law for the establishment of a Utilities Commission to take over the waterworks sewage disposal plant and the hydro system.

The Canada Steamship Line, Montreal staff have enlisted for overseas service. The eligible men are to be enlisted in one battalion, and officers, who are to go to Halifax at once to qualify for commissions, will lead the company. There are forty-six Canada Steamship Line in the trenches now.

Kingston, Ont.—The announcement of United States Consul F. F. Johnson last Monday shows the export trade from this port to the States for the year 1915 to have reached the million dollar mark, giving Kingston third place in the Dominion, with Toronto first and Ottawa second.

Toronto, Ont.—It is reported from La Porte, Ind., that the plants of the M. Rumley Co., and Rumley Products Co., capitalized at $32,000,000, were bought in at receiver's sale for $4,000,000 on Dec. 9 by reorganized Rumley companies. The Rumley Co. have a branch in this city.

Toronto, Ont.—Roads to the extent of 250 miles have been permanently improved and macadamized in various parts of the Province of Ontario this year. On colonization roads $220,078.-19 has been spent. The total cost of the Hamilton-Toronto highway will be about $900,000. The estimated cost was $600,000.

Montreal, Que.—The shareholders of the Canada Cement Company, at a special meeting recently unanimously authorized an application for supplementary letters patent permitting the company to embark on the manufacture of war munitions. General Manager Jones gave an optimistic forecast of the results expected from the business, and the shareholders were informed that no new financing would be necessary.

Montreal, Que.—Three departments of the Grand Trunk Railway shops at Point St. Charles were gutted by fire which broke out at noon on Dec. 10 in the tube

shop and spread to the blacksmith shop and the erecting plant. The fire was caused by a tube striking and breaking an oil pipe, the oil igniting and flowing in all directions. The tube shop, blacksmith shop and part of the erecting shop were destroyed.

Toronto Harbor Works.—It is announced that the Toronto harbor work in the vicinity of Cherry Street is to proceed as weather permits. The Canadian Stewart Co., the contractors, have undertaken to make good the defective construction by a sub-contractor disclosed upon examination by Government experts, and the work will proceed without further hitch in favorable weather. The restoration of the imperfect construction will be done without cost to the Government, it is said. The contractors say there will be no unnecessary delay.

Grain Moving on N. T. R.—Arrangements have now been completed for the official inspection, weighing and grading of grain at Winnipeg so that grain consigned by the N. T. R. will not have to go first to Fort William or Port Arthur. At present all certificates are granted at the head of the lakes. At Halifax, where most of the grain handled will be shipped, Mr. Cochrane has made arrangements for the storage of some five hundred cars, so that the grain will be protected until shipped. There is no elevator accommodation at Halifax. The Government expects to get a large share of the winter all-rail shipments.

National Steel Car Co.—In a letter to the shareholders, Vice-President Basil Magor states that the company has orders aggregating $8,000,000 for rolling stock to be used on foreign railways, as well as other large contracts, and in addition the company is building steel sleeping coaches, first-class coaches and motor trucks. Mr. Magor states that the fiscal year ending on November 30 will show approximately $450,000 of profits, and that, therefore, if the dividend liability is removed without depleting too much the working capital, the company's position will be substantially improved.

Personal

Frederick Powell, president of the Rideau Lumber Co., Ottawa, Ont., died on December 3.

W. G. Worden, a graduate of Toronto University, has been appointed town engineer of Oshawa, Ont., to succeed F. A. Chappell, who has resigned.

E. M. Smith, president of the American Shipbuilding Co., who owns the dry dock at Port Arthur, Ont., died suddenly at Buffalo, N.Y., on Dec. 8.

A. G. **McAvity**, manager of the Canadian Blower & Forge Co., Berlin, Ont., has been engaged by the Imperial Munitions Board in an advisory capacity.

Lieut. F. Y. **Harcourt**, who for several years has been district harbor engineer at Port Arthur, Ont., has received a commission in the 94th Battalion of the C.E.F.

W. L. **Helliwell**, manager of the Gurney Northwest Foundry Co., Winnipeg, Man., has been appointed to a more important position on the staff of the firm at the head office in Montreal.

Thomas J .**Walsh**, chief operating engineer of the high-level pumping station, Toronto, Ont., who had been in the civic service for a long number of years, died on Dec. 12 at his home, 40 Rathnally Avenue, in his 55th year.

H .S. **Carmichael** has been appointed freight and passenger agent at London, England, of the Canadian Pacific Ocean Services. He completes twenty years' service with the C. P. R. this week. H. S. Dring succeeds him as passenger agent of the C. P. R.

L. C. **Ord**, assistant works manager, Angus car shops, C. P. R., Montreal, Que., has been appointed Lieutenant in No. 1 Overseas Battery of the Siege Artillery, C.E.F., and has been granted leave of absence for active service. Lieutenant Ord sailed from Halifax for the front November 22nd.

F. **Perry** has been appointed a member of the Imperial Munitions Board. Mr. Perry has for some years represented in Canada a large private banking house of London. At one time he was attached to the British Colonial Office, and later was Imperial secretary to Lord Milner in South Africa and chairman of a native labor association in South Africa.

New Incorporations

The York Paper Box Co. has been incorporated at Toronto, with a capital of $40,000, to manufacture all kinds of paper boxes, etc., at Toronto. Incorporators: James Douglas Woods and William Batten McPherson, of Toronto.

The Chemical Refinery, Ltd., has been incorporated at Ottawa, with a capital of $100,000, to manufacture drugs, chemicals and fertilizers at St. Catharines, Ont. Incorporators: John P. Mitchell, Harry W. Page and W. P. Crow, all of Toronto.

The J. S. Lewis & Sons, Ltd., has been incorporated at Ottawa, with a capital of $100,000, to carry on the business of lumbering and manufacturing of wood

and other materials at Truro, N.S. Incorporators: George Ezra Morton Lewis and Frank Leslie Lewis, of Truro, N.S.

Railways - Bridges

Brantford, Ont.—It is understood that arrangements are practically completed for the creation of a Union Radial Station in this city to be used jointly by the Brantford and Hamilton and Lake Erie and Northern Railways. The intention is to place it on the old site of the Brantford Ice Co., and it will cost approximately $30,000.

Berlin, Ont.—By a vote of eight to four the City Council decided on Dec. 6 to submit a hydro-radial by-law for $779,000 to the ratepayers. Waterloo Town and Waterloo Township Councils also voted in favor of submitting similar by-laws, the former for $193,000, and the latter for $521,903. Engineer F. A. Gaby, of the Ontario Hydro-Electric Commission, attended all three meetings. The route through Berlin agreed upon by the citizens was not defined in the by-law.

Building Notes

Toronto, Ont.—A building permit has been issued to W. H. Harris, in trust, for the erection of a two-storey brick mill construction warehouse on the north side of Richmond Street, near John Street, costing $18,000.

Wood-Working

Calgary, Alta.—The Rumley Co. propose building a warehouse here and make the city a distributing centre.

Hamilton, Ont.—Spontaneous combustion started a blaze in the Semmens & Evel casket factory on December 7, and did $2,000 damage.

Deseronto, Ont.—The Canada Hardwood Mfg. Co. is in the market for sawmill and wood-working machinery, auto-

Marine

Port Arthur, Ont.—The Western Dry Dock & Shipbuilding Co.'s plant will be working to capacity during the coming winter months. The company have on hand several freighters for lake service.

Port Arthur, Ont.—Freight rates from Port Arthur to Buffalo have risen from a rate of from three and a half to four cents a bushel, which prevailed during the regular season, to five and a half cents.

North Vancouver, B.C.—The City Council will immediately take steps to offer some inducements for the establishment of a drydock on the North Shore. The Board of Trade will support the scheme.

John L. Nelson, who came to the Province of British Columbia in March, 1913, as superintendent of dredging, has sent in his resignation to the Department of Public Works.

Port Arthur, Ont.—The Canadian steamer, W. Grant Morden, the longest boat on the lakes, left on Dec. 9 for Port McNicoll with 760,000 bushels of oats, the largest cargo ever taken from this port, being equivalent to 400 loaded cars.

Brockville, Ont.—The Department of Marine and Fisheries has awarded the contract for the erection of a lighthouse one the river front at Fulford's Point, west of here. The building will be twenty feet high with concrete foundation.

Collingwood, Ont.—The steamer Hamonic of the Northern Navigation Co.'s Lake Superior fleet is coming to Collingwood about Dec. 15th and will probably remain throughout the winter. The object of the visit is to go to the dry dock for repairs.

Catalogues

Milling Machines.—A bulletin issued by the Fox Machine Co., Grand Rapids, Mich., describes and illustrates the "Fox" No. 3 hand and power feed milling machine. A specification is included giving the principal dimensions and shipping weights, etc.

Air Heater and Air Blower.—Bulletin No. 219, issued by the B. F. Sturtevant Co., Boston, Mass., deals with the "Sturtevant" electric air heater and blower. The apparatus is fully described and the various purposes for which it can be used are given in detail. A list of sizes with capacities is included and the illustrations show a few different types indicating their varied application.

Friction Clutches made by the Carlyle, Johnson Machine Co., Manchester, Conn., are illustrated and described in catalogue E, recently issued. The various classes of work for which this clutch is suitable are described separately with illustrations showing the mountings. Tables are included giving the principal dimensions, accompanied by diagrams showing to what part the dimensions refer. Other equipment for use with these clutches is also described and illustrated.

Oxy-Acetylene Apparatus.—The L'Air Liquide Society of Montreal, Que., have issued a folder containing a number of leaflets describing their oxy-acetylene apparatus for welding and cutting. The many types of blowpipes for various purposes are described fully and tables give the capacity for each size. Other leaflets describe and illustrate acetylene generators. A number of welding supplies and materials are dealt with in detail, accompanied by price lists. The leaflets are fully illustrated and included in views of the plants at work.

The Phenix Force Feed Lubricator, Bulletin No. 50, recently issued by the Richardson-Phenix Co. of Milwaukee, Wis., contains a complete description of the Phenix ratchet type lubricator; it also describes a new type, known as the model "T." The catalogue contains an interesting diagram, giving a comparison of the way in which mechanical and hydrostatic lubricators feed oil to engine cylinders. Numerous illustrations show the application of the Phenix lubricator to different types of engines, pumps, steam hammers, etc.

Establishing and Maintaining Boiler Room Economy is the title of a paper presented before the Ohio Society of Mechanical, Electrical and Steam Engineers, by Geo. H. Gibson, developing the thesis that the most important requisites to further improvements of boiler plant economy are means of recording boiler performance—that is, of determining the number of pounds of water evaporated per pound of coal. Information obtained sporadically, as by means of short boiler tests, is not so suitable for this purpose as is information supplied continuously, as by a feed water meter. The relative values of different kinds of coal, the improvement in evaporation following cleaning of soot and scale off heating surfaces, or the stopping up of air leaks, the relative merits of different methods of firing, are then readily demonstrated, and scientific management becomes easy and natural. That is, the ways and means of attaining a certain standard of performance having been demonstrated, the management is in a position to ask for good results continuously; in fact, standard rules of operation—that is, directions as to methods of handling fires, regulation of draft, blowing of soot, banking fires, carrying over-loads, etc., can be written out, so that any man following instructions can obtain good results. The installation further arouses the pride of the engineer, and makes it possible to reward skill or attention to duty. This pamphlet is being distributed by the Harrison Safety Boiler Works, Philadelphia, Pa.

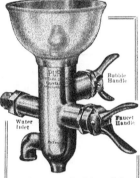

If what you want is not advertised in this issue consult the Buyers' Directory at the back.

The advertiser would like to know where you saw his advertisement—tell him.

If what you want is not advertised in this issue consult the Buyers' Directory at the back.

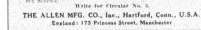

If what you want is not advertised in this issue consult the Buyers' Directory at the back.

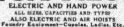

CANADIAN MACHINERY BUYERS' DIRECTORY

TO OUR READERS—Use this directory when seeking to buy any machinery or power equipment.
You will often get information that will save you money.

TO OUR ADVERTISERS—Send in your name for insertion under the headings of the lines you
make or sell.

TO NON-ADVERTISERS—A rate of $5 per line a year is charged non-advertisers.

Abrasive Materials.
Can. Fairbanks-Morse Co., Montreal.
Carborundum Co., Niagara Falls, N.Y.
Ford-Smith Machine Co., Hamilton, Ont.
Gardner Machine Co., Beloit, Wis.
Norton Co., Worcester, Mass.
H. W. Petrie, Toronto.
Stevens, F. B., Detroit, Mich.

Acetylene.
L'Air Liquide Society, Montreal, Toronto.
LeVer Bros., Toronto.

Acetylene Generators.
L'Air Liquide Society, Montreal, Toronto.
LeVer Bros., Toronto.

Accumulators, Hydraulic.
Can. Boomer & Boschert Press Co., Montreal.
Charles F. Elmes Eng. Wks., Chicago.
Mesta Machine Co., Pittsburg, Pa.
William R. Perrin, Ltd., Toronto.
Southwark Foundry & Machine Co., Philadelphia.
Wm. Tod Company, Youngstown, O.
Watson-Stillman Co., Aldene, N.J.
Wood, R. D., & Co., Philadelphia.

Air Compressors.
Canadian Ingersoll-Rand Co., Ltd., Montreal.
Cleveland Pneumatic Tool Co. of Canada, Toronto.
Curtis Pneumatic Machinery Co., St. Louis, Mo.
H. W. Petrie, Toronto.
Smart-Turner Machine Co., Hamilton.

Air Hoists.
Northern Crane Works, Ltd., Walkerville, Ont.
Whiting Foundry Equipment Co., Harvey, Ill.

Air Hose.
Can. H. W. Johns-Manville Co., Limited, Toronto.
Cleveland Pneumatic Tool Co. of Canada, Toronto.
Can. Ingersoll-Rand Co., Montreal.

Air Receivers.
Buffalo Forge Co., Buffalo, N.Y.
Can. Sirocco Co., Ltd., Windsor, Ont.

Air Washers.
Can. H. W. Johns-Manville Co., Limited, Toronto.

Ammeters.
Lints-Porter Co., Toronto.

Aluminium.
Tallman Brass & Metal Co., Hamilton.

Alloys, Steel.
H. A. Drury Co., Ltd., Montreal.
Hawkridge Bros. Co., Boston, Mass.
Vanadium Alloys Steel Co., Pittsburg, Pa.
Vulcan Crucible Steel Co., Aliquippa, Pa.

Annunciator Systems.
Lints-Porter Co., Toronto.

Arbors.
Can. Fairbanks-Morse Co., Montreal.
Cleveland Twist Drill Co., Cleveland.
Morse Twist Drill and Machine Co., New Bedford.
H. W. Petrie, Toronto.
Pleasistville Foundry, Pleasistville, Que.
Pratt & Whitney Co., Dundas, Ont.

Assembling Stands.
Skinner Chuck Co., New Britain, Conn.

Automatic Chucks.
Garvin Machine Co., New York.

Asbestos Packing.
Can. H. W. Johns-Manville Co., Limited, Toronto.

Autogenous Welding and Cutting Plants.
L'Air Liquide Society, Montreal, Toronto.
LeVer Bros., Toronto.

Automatic Index Milling.
Garvin Machine Co., New York.
National Machinery and Supply Co., Hamilton.
H. W. Petrie, Toronto.

Automatic Machinery.
Baird Machine Co., Bridgeport, Conn.
A. R. Williams Machy. Co., Toronto.
Gardner, Robt., & Son, Montreal.
Girard Machine & Tool Co., Philadelphia, Pa.
Motch & Merryweather Machy. Co., Cleveland, O.
National Machinery & Supply Co., Hamilton.
H. W. Petrie, Toronto.
Pratt & Whitney Co., Dundas, Ont.
Owen Sound Iron Works Co., Owen Sound.
Windsor Machine Co., Windsor, Vt.

Automatic Multiple Spindle.
Windsor Machine Co., Windsor, Vt.

Automatic Wood Screw Machines.
Asa F. Cook Co.

Axle Cutters.
Butterfield & Co., Rock Island, Que.
A. B. Jardine & Co., Hespeler, Ont.

Automobile Fenders.
Dominion Forge & Stamping Co., Walkerville, Ont.

Automobile Hoods.
Dominion Forge & Stamping Co., Walkerville, Ont.

Babbitt Metal.
Can. Fairbanks-Morse Co., Montreal.
Hoyt Metal Co., Toronto.
Magnolia Metal Co., Montreal.
H. W. Petrie, Toronto.
Tallman Brass & Metal Co., Hamilton.

Baking Ovens.
Owen Equipment & Mfg. Co., New Haven, Conn.
Owen Sound Iron Works Co., Owen Sound.

Ball Bearings.
Can. Fairbanks-Morse Co., Montreal.
Chapman Double Ball Bearing Company, Toronto.
H. W. Petrie, Toronto.

Barrenshing Machines.
Baird Machine Co., Bridgeport, Conn.

Banding Machines, Hydraulic.
West Tire Setter Co., Rochester, N.Y.

Barrels, Steel Shop.
Baird Machine Co., Bridgeport, Conn.
Cleveland Wire Spring Co., Cleveland.

Bar Steel.
Steel Co. of Canada, Hamilton, Ont.

Bars, Boring.
Charles F. Elmes Eng. Works, Chicago.
Niles-Bement-Pond Co., New York.
Owen Sound Iron Works Co., Owen Sound.

Bar Benders and Straight Edges.
Steel Bending Brake Works, Ltd., Chatham, Ont.

Bar Benders, Hydraulic.
Charles F. Elmes Eng. Works, Chicago
Watson-Stillman Co., Aldene, N.J.

Bar Twisting Machines.
Mesta Machine Co., Pittsburg, Pa.

Bafleries and Accessories.
Lints-Porter Co., Toronto.

Bell Systems.
Lints-Porter Co., Toronto.

Belt Benches.
Tabor Mfg. Co., Philadelphia, Pa.

Belting, Balata.
F. Reddaway & Co., Montreal.

Belting, Camel Hair.
F. Reddaway & Co., Montreal.

Belt Dressing and Cement.
Graton & Knight Mfg. Co., Montreal.
Belt Fasteners.
F. Reddaway & Co., Montreal.

Belt Lacing, Leather.
Graton & Knight Mfg. Co., Montreal.

Belting, Chain.
Can. Fairbanks-Morse Co., Montreal.
Graton & Knight Mfg. Co., Montreal.
Jones & Glassco, Montreal.
Morse Chain Co., Ithaca, N.Y.
H. W. Petrie, Toronto.

Belting, Cotton
General Supply Co. of Canada. Ltd., Ottawa.
Dominion Belting Co., Hamilton.
H. W. Petrie, Toronto.
F. Reddaway & Co., Montreal.

Belting Leather.
Can. Fairbanks-Morse Co., Montreal.
General Supply Co. of Canada, Ltd., Ottawa.
Girard Machine & Tool Co., Philadelphia, Pa.
Graton & Knight Mfg. Co., Montreal.
Main Belting Co., Montreal.
Morse Chain Co., Ithaca, N.Y.
H. W. Petrie, Toronto.

Belting, stitched Cotton Duck.
General Supply Co. of Canada, Ltd., Ottawa.
Dominion Belting Co., Hamilton, Ont.
Main Belting Co., Montreal.
H. W. Petrie, Toronto.
F. Reddaway & Co., Montreal.

Belting, Rubber.
Can. H. W. Johns-Manville Co., Limited, Toronto.

Benders, Angle and Tee Iron.
Can. Buffalo Forge Co., Montreal.
Watson-Stillman Co., Aldene, N.J.

Bending Machinery.
John Bertram & Sons Co., Dundas.
Bertrams, Limited, Edinburgh, Scotland.
Bliss, E. W., Co., Brooklyn, N.Y.

Brown, Boggs Co., Ltd., Hamilton, Canada.
Can. Buffalo Forge Co., Montreal.
Can. Machinery Corporation, Galt, Ont.
Charles F. Elmes Eng. Works, Chicago
Jardine, A. B., & Co., Hespeler, Ont.
National Machinery Co., Tiffin, Ohio.
National Machinery & Supply Co., Hamilton.
Niles-Bement-Pond Co., New York.
Owen Sound Iron Works Co., Owen Sound.
H. W. Petrie, Toronto.
Toledo Machine & Tool Co., Toledo, O.
Steel Bending Brake Works, Chatham, Ont.
Watson-Stillman Co., Aldene, N.J.

Bins, Steel.
Dennis Wire & Iron Works Co., Ltd., London, Canada.
MacKinnon, Holmes Co., Sherbrooke.
Toronto Iron Works, Ltd., Toronto.

Bit Brace Tools.
Wells Bros. Co., Greenfield, Mass.
Wilt Twist Drill Co. of Canada, Ltd., Walkerville, Ont.

Blast Gauges, Cupola.
Can. Buffalo Forge Co., Montreal.
Sheldons, Ltd., Galt, Ont.
Whiting Foundry Equipment Co., Harvey, Ill.

Blocks, Lifting.
Northern Crane Works, Walkerville.

Blowers.
Can. Buffalo Forge Co., Montreal.
Can. Sirocco Co., Ltd., Windsor, Ont.
Chicago Flexible Shaft Co., Chicago.
Girard Machine & Tool Co., Philadelphia, Pa.
Sheldons, Ltd., Galt, Ont.
Southwark Foundry & Machine Co., Philadelphia.

Blow Pipes and Regulators.
L'Air Liquide Society, Montreal, Toronto.
LeVer Bros., Toronto.

Bluing Ovens.
Owen Equipment & Mfg. Co., New Haven, Conn.

Boards.
Geo. A. Joyce Co., Ltd., New York, N.Y.

Boilers.
Can. Locomotive Co., Kingston, Ont.
General Supply Co. of Canada, Ltd., Ottawa.
MacKinnon, Holmes Co., Sherbrooke, Que.
National Machinery & Supply Co., Hamilton.
Owen Sound Iron Works Co., Owen Sound.
H. W. Petrie, Toronto.
Pleasistville Foundry, Pleasistville, Que.

Boiler Compounds.
Can. H. W. Johns-Manville Co., Limited, Toronto.

Boiler Graphite.
Dixon Crucible Co., Jersey City, N.J.

Boiler Makers' Supplies.
Jno. F. Allen Co., New York.

Bolt Cutters and Nut Tappers.
Wells Brothers Co., Greenfield, Mass.

Bolts.
Galt Machine Screw Co., Galt, Ont.
London Bolt & Hinge Works, London, Ont.
Steel Co. of Canada, Hamilton, Ont.

Bolt and Nut Machinery.
A. R. Williams Machy. Co., Toronto.
John Bertram & Sons Co., Dundas.
Owen Sound Iron Works Co., Owen Sound.
Gardner, Robt., & Son, Montreal.
Landis Machine Co., Waynesboro, Pa.
National Machinery Co., Tiffin, O.
National Machinery & Supply Co., Hamilton.
H. W. Petrie, Toronto.
Wiley & Russell Co., Greenfield, Mass.

Boring Machines, Upright and Horizontal.
John Bertram & Sons Co., Dundas.
Colburn Machine Tool Co., Franklin, Pa.
Garlock-Machinery, Toronto.
Girard Machine & Tool Co., Philadelphia, Pa.
Hill, Clarke & Co., of Chicago, Chicago, Ill.
Motch & Merryweather Machy. Co., Cleveland, O.
National Machinery & Supply Co., Hamilton.
Niles-Bement-Pond Co., New York.
Oliver Machy. Co., Grand Rapids.
Slow Mfg. Co., Binghamton, N.Y.

Boring Machines, Pneumatic Cylinder.
Baker Brothers, Toledo, O.

Cleveland Pneumatic Tool Co. of Canada, Toronto.
Can. Fairbanks-Morse Co., Montreal.
Can. Ingersoll-Rand Co., Montreal.
Independent Pneumatic Tool Co., Chicago, Ill.
H. W. Petrie, Toronto.
Slow Mfg. Co., Binghamton, N.Y.

Boring and Turning Mills.
John Bertram & Sons Co., Dundas.
Girard Machine & Tool Co., Philadelphia, Pa.
National Machinery & Supply Co., Hamilton.
Niles-Bement-Pond Co., New York.
H. W. Petrie, Toronto.

Boxes, Annealing, Charging.
Mesta Machine Co., Pittsburg, Pa.
Box Puller.
Jardine, A. B., & Co., Hespeler, Ont.

Boxes, Steel Shop.
Cleveland Wire Spring Co., Cleveland.

Boxes, Tote.
Cleveland Wire Spring Co., Cleveland.

Brakes.
Brown, Boggs & Co., Hamilton, Can.
Whiting Foundry Equipment Co., Harvey, Ill.

Brakes, Heavy Plate Bending and Cornice.
Steel Bending Brake Works, Ltd., Chatham, Ont.

Brass Working Machinery.
A. R. Williams Machy. Co., Toronto.
Gardner, Robt., & Son, Montreal.
Girard Machine & Tool Co., Philadelphia, Pa.
National Machinery & Supply Co., Hamilton.
Warner & Swasey Co., Cleveland.
Niles-Bement-Pond Co., New York.
H. W. Petrie, Toronto.

Brick Cars.
Can. Buffalo Forge Co., Montreal.
Sheldons, Ltd., Galt, Ont.

Brick Dryers.
Can. Buffalo Forge Co., Montreal.
Can. Sirocco Co., Ltd., Windsor, Ont.
Sheldons, Ltd., Galt, Ont.

Brick Machinery.
Eastern Machinery Co., New Haven.
Sheldons, Ltd., Galt, Ont.

Bridges, Railway and Highway.
Can. Bridge Co., Walkerville, Ont.
MacKinnon, Holmes Co., Sherbrooke.

Bubblers.
Puro Sanitary Drinking Fountain Co., Haydenville, Mass.

Buckets, Clam Shell, Crab and Dump.
Northern Crane Works, Ltd., Walkerville, Ont.
Whiting Foundry Equipment Co., Harvey, Ill.

Buffing and Polishing Machinery.
Canadian Hart Wheels, Ltd., Hamilton, Ont.
Ford-Smith Machine Co., Hamilton, Ont.
Girard Machine & Tool Co., Philadelphia, Pa.
New Britain Machine Co. New Britain, Conn.

Bulldozers.
John Bertram & Sons Co., Dundas.
E. W. Bliss Co., Brooklyn, N.Y.
Canada Mach. Corporation, Galt, Ont.
National Machinery & Supply Co., Hamilton, Ont.
Watson-Stillman Co., Aldene, N.J.

Burners, Enclosed Flame Gas.
Owen Equipment & Mfg. Co., New Haven, Conn.

Burners, Fuel, Oil and Natural Gas.
Northern Crane Works, Ltd., Walkerville, Ont.
Whiting Foundry Equipment Co., Harvey, Ill.

Burring Reamers.
Wells Brothers Company, Greenfield Mass.
Wilt Twist Drill Co. of Canada, Ltd., Walkerville, Ont.

Butteries.
Wells Brothers Company, Greenfield Mass.

Burrs, Iron and Copper.
Parmenter & Bulloch Co., Gananoque
Canners' Machinery.
Bliss, E. W., Co., Brooklyn, N.Y.
Brown, Boggs & Co., Hamilton, Can.
National Machinery & Supply Co., Hamilton, Ont.

Caissons.
Toronto Iron Works, Ltd., Toronto.

Cars, Charging Box Ingot.
Mesta Machine Co., Pittsburg, Pa.

Cars, Industrial.
Can. Buffalo Forge Co., Montreal.
Can. Fairbanks-Morse Co., Montreal.
Sheldons, Limited, Galt, Ont.
Whiting Foundry Equipment Co., Harvey, Ill.

Castings, Aluminum.
Cunningham & Son, St. Catharines, Ont.
Owen Sound Iron Works Co., Ltd., Owen Sound, Ont.
St. Lawrence Foundry, Galt, Ont.
Tallman Brass & Metal Co., Hamilton

Castings, Air Furnaces.
Wm. Tod Company, Youngstown, O.

Castings, Brass.
Cunningham & Son, St. Catharines, Ont.
Alexander Fleck, Ltd., Ottawa.
T. C. Lawrence Foundry, Galt, Ont.
Mesta Machine Co., Pittsburg, Pa.
Owen Sound Iron Works Co., Owen Sound.
Plessisville Foundry, Plessisville, Que.
Tallman Brass & Metal Co., Hamilton
Wm. Tod Company, Youngstown, O.

Castings, Bronze.
Cunningham & Son, St. Catharines, Ont.
Mesta Machine Co., Pittsburg, Pa.
Tallman Brass & Metal Co., Hamilton
Wm. Tod Company, Youngstown, O.

Castings, Copper.
Cunningham & Son, St. Catharines, Ont.
Tallman Brass & Metal Co., Hamilton, Ont.

Castings, Gray Iron.
Brown, Boggs Co., Ltd., Hamilton, Canada.
Erie Foundry Co., Erie, Pa.
Alexander Fleck, Ltd., Ottawa.
Gardner, Robt., & Son, Montreal.
Hull Iron & Steel Foundries, Ltd., Hull, Quebec.
Wm. Kennedy & Sons, Ltd., Owen Sound.
Mesta Machine Co., Pittsburg, Pa.
Owen Sound Iron Works Co., Owen Sound.
Plessisville Foundry, Plessisville, Que.
Wm Tod Company, Youngstown, O.

Castings, Steel Chrome and Manganese Steel.
Hull Iron & Steel Foundries, Ltd., Hull, Quebec.
Wm. Kennedy & Sons, Ltd., Owen Sound, Ont.
Mesta Machine Co., Pittsburg, Pa.
Wm. Tod Company, Youngstown, O.

Castings, Malleable.
Galt Malleable Iron Co., Galt.

Castings, Nickel Steel.
Hull Iron & Steel Foundries, Ltd., Hull, Quebec.
Mesta Machine Co., Pittsburg, Pa.

Cement, Disc Wheel.
Gardner Machine Co., Beloit, Wis.

Cement, Iron.
Can. H. W. Johns-Manville Co., Limited, Toronto.
Shelton Metallic Filler Co., Derby, O.

Cement Machinery.
Can. Fairbanks-Morse Co., Montreal.
Gardner, Robt., & Son, Montreal.
National Machinery & Supply Co., Hamilton, Ont.
Owen Sound Iron Works Co., Owen Sound.
H. W. Petrie, Toronto.

Centre Reamers.
Wells Brothers Co., Greenfield, Mass.

Centering Machines.
John Bertram & Sons Co., Dundas.
Gardner, Robt., & Son, Montreal.
Girard Machine & Tool Co., Philadelphia, Pa.
Hurlbut, Rogers Machinery Co., South Sudbury, Mass.
National Machinery & Supply Co., Hamilton.
Niles-Bement-Pond Co., New York.
Pratt & Whitney Co., Dundas, Ont.

Centrifugal Pumps.
Can. Buffalo Forge Co., Montreal.
H. W. Petrie, Toronto.
Pratt & Whitney Co., Dundas, Ont.
Southwark Foundry & Machine Co., Philadelphia, Pa.
Smart-Turner Machine Co., Hamilton, Ont.

Chain Blocks.
Can. Fairbanks-Morse Co., Montreal.
National Machinery & Supply Co., Hamilton.
H. W. Petrie, Toronto.

Chains, Silent and Transmission.
Jones & Glassco, Montreal.
Morse Chain Co., Ithaca, N.Y.
Plessisville Foundry, Plessisville, Que.

Chemists.
Can. Inspection & Testing Laboratories, Ltd., Montreal.
Toronto Testing Laboratory, Ltd., Toronto.

Chucks, Aero, Automatic.
Garvin Machine Co., New York.

Chucks, Drill, Lathe and Universal.
John Bertram & Sons Co., Dundas, Ont.
Buffalo Forge Co., Buffalo, N.Y.
Can. Fairbanks-Morse Co., Montreal.
Cleveland Twist Drill Co., Cleveland.
Cushman Chuck Co., Hartford, Conn.
Gardner, Robt., & Son, Montreal.
Girard Machine & Tool Co., Philadelphia, Pa.
Wells Brothers Co., Greenfield, Mass.
Jacobs Mfg. Co., Hartford, Conn.
Ker & Goodwin, Brantford.
Modern Tool Co., Erie, Pa.
Morse Twist Drill & Machine Co., New Bedford.
National Machinery & Supply Co., Hamilton.
H. W. Petrie, Toronto.
Skinner Chuck Co., New Britain, Conn.
D. E. Whiton Machine Co., New London, Conn.
Wilt Twist Drill Co. of Canada, Ltd., Walkerville, Ont.

Chucks, Drill, Automatic and Keyless.
Buffalo Forge Co., Buffalo, N.Y.

Chucks, Ring Wheel.
Gardner Machine Co., Beloit, Wis.

Chucking Machines.
Garvin Machine Co., New York.
Girard Machine & Tool Co., Philadelphia, Pa.
New Britain Machine Co., New Britain, Conn.
Niles-Bement-Pond Co., New York.
Turner Machine Co., Danbury, Conn.
Warner & Swasey Co., Cleveland, O.

Circular Safety Cylinders.
Oliver Machy. Co., Grand Rapids, Mich.

Clocks, Time and Watchman's.
Lintz-Porter Co., Toronto.

Cloth and Wool Dryers.
Canada Wire & Iron Goods Co., Hamilton, Ont.
Sheldons, Limited, Galt.

Clutches.
Eastern Machinery Co., New Haven, Conn.
Jones & Glassco, Montreal.
Owen Sound Iron Works Co., Owen Sound.
Positive Clutch & Pulley Works, Ltd., Toronto.

Coal Handling Machinery.
Northern Crane Works, Ltd., Walkerville, Ont.
Whiting Foundry Equipment Co., Harvey, Ill.

Coke and Coal.
Hanna & Co., M. A., Cleveland, O.

Collectors, Pneumatic.
Can. Buffalo Forge Co., Montreal.
Sheldons, Limited, Galt.

Compressors, Air.
Cleveland Pneumatic Tool Co. of Canada, Toronto.
Independent Pneumatic Tool Co., Chicago.
Mesta Machine Co., Pittsburg, Pa.
National Machinery & Supply Co., Hamilton.
H. W. Petrie, Toronto.
Southwark Foundry & Machine Co., Philadelphia, Pa.
The Smart-Turner Machine Co., Hamilton.

Concentrating Plant.
Gardner, Robt., & Son, Montreal.

Concrete Mixers.
A. R. Williams Machy. Co., Toronto.
Can. Fairbanks-Morse Co., Montreal.
National Machinery & Supply Co., Hamilton.
H. W. Petrie, Toronto.

Concrete Reinforcement.
Canada Wire Goods Mfg. Co., Hamilton.

Condensers.
Can. Buffalo Forge Co., Montreal.
Mesta Machine Co., Pittsburg, Pa.
The Smart-Turner Machine Co., Hamilton.
Southwark Foundry & Machine Co., Philadelphia.
Wm. Tod Company, Youngstown, O.

Contracting Engineers, Electrical
Lintz-Porter Co., Toronto.

Controllers and Starters, Electric Motor.
A. R. Williams Machy. Co., Toronto.
H. W. Petrie, Toronto.
Toronto & Hamilton Electric Co., Hamilton, Ont.

Conveyor Machinery.
Beath, W. D., & Son, Toronto.
Can. Fairbanks-Morse Co., Montreal.
National Machinery & Supply Co., Hamilton, Ont.
H. W. Petrie, Toronto.
Plessisville Foundry, Plessisville, Que.
The Smart-Turner Machine Co., Hamilton.

Coping Machines.
Can. Buffalo Forge Co., Montreal.
John Bertram & Sons Co., Dundas.
National Machinery & Supply Co., Hamilton, Ont.
Niles-Bement-Pond Co., New York.

Cornice Brakes.
Brown Boggs Co., Ltd., Hamilton, Canada.
Steel Bending Brake Wks., Chatham.

Counting Machines.
W. N. Durant Co., Milwaukee, Wis.
National Scale Co., Chicopee Falls, Mass.
C. J. Root Co., Bristol, Conn.

Counterbores and Countersinks.
Cleveland Twist Drill Co., Cleveland.
Morse Twist Drill & Machine Co., New Bedford.
Pratt & Whitney Co., Dundas, Ont.
Wells Bros. Co., Greenfield, Mass.
Wilt Twist Drill Co. of Canada, Ltd., Walkerville, Ont.

Countershafts.
Baird Machine Co., Bridgeport, Conn.
Wells Bros. Co., Greenfield, Mass.

Country House Lighting and Cooking.
Can. Blaugas Co., Montreal.

Couplings.
Can. H. W. Johns-Manville Co., Ltd., Toronto.
Eastern Machinery Co., New Haven, Conn.
Gardner, Robt., & Son, Montreal.
Owen Sound Iron Works Co., Owen Sound, Ont.

Couplings, Air Hose.
Cleveland Pneumatic Tool Co. of Canada, Toronto.

Crabs, Travelling.
Owen Sound Iron Works Co., Owen Sound.

Cranes, Locomotive.
Northern Crane Works, Walkerville.

Cranes, Gantry.
Northern Crane Works, Walkerville.
Smart-Turner Machine Co., Hamilton, Ont.
Whiting Foundry Equipment Co., Harvey, Ill.

Cranes, Goliath.
Herbert Morris Crane & Hoist Co., Ltd., Toronto.
Northern Crane Works, Walkerville.
Whiting Foundry Equipment Co., Harvey, Ill.

Cranes, Hydraulic.
Southwark Foundry & Machine Co., Philadelphia.
Watson-Stillman Co., Aldene, N.J.

Cranes, Pneumatic.
Northern Crane Works, Walkerville.
Whiting Foundry Equipment Co., Harvey, Ill.

Cranes, Post Jib.
Northern Crane Works, Walkerville.
Smart-Turner Machine Co., Hamilton, Ont.
Whiting Foundry Equipment Co., Harvey, Ill.

Cranes, Portable.
Northern Crane Works, Walkerville.
Whiting Foundry Equipment Co., Harvey, Ill.

Cranes, Swing Jib.
Northern Crane Works, Walkerville.
Smart-Turner Machine Co., Hamilton, Ont.
Whiting Foundry Equipment Co., Harvey, Ill.

Cranes, Transfer.
Northern Crane Works, Walkerville.
Smart-Turner Machine Co., Hamilton, Ont.
Whiting Foundry Equipment Co., Harvey, Ill.

Cranes, Wall.
Northern Crane Works, Walkerville.
Smart-Turner Machine Co., Hamilton, Ont.
Whiting Foundry Equipment Co., Harvey, Ill.

Cranes, Travelling Electric and Hand Power.
Dominion Bridge Co., Montreal.
Niles-Bement-Pond Co., New York.
Northern Crane Works, Walkerville.
Whiting Foundry Equipment Co., Harvey, Ill.

Cranes, All Kinds.
Northern Crane Works, Walkerville.
Owen Sound Iron Works Co., Owen Sound, Ont.
Southwork Foundry & Machine Co., Philadelphia.
Harvey, Ill.
Whiting Foundry Equipment Co.

Crank Pin Turning Machine.
Niles-Bement-Pond Co., New York.

Crimps, Leather.
Graton & Knight Mfg. Co., Montreal.

Cupolas.
Can. Buffalo Forge Co., Montreal.
Northern Crane Works, Walkerville.
H. W. Petrie, Toronto.
Sheldons, Ltd., Galt, Ont.
Whiting Foundry Equipment Co., Harvey, Ill.

Cupola and Blast Gate Blowers.
Can. Sirocco Co., Ltd., Windsor, Ont.

Cupola Blast Gauges & Blowers.
Sheldons, Ltd., Galt, Ont.

Cutters, Angle, Tee Iron and Bar.
Can. Buffalo Forge Co., Montreal.

Cutters, Flue.
Independent Pneumatic Tool Co., Chicago.
Cleveland Pneumatic Tool Co. of Canada, Toronto.

Cutters, Pipe.
Can. Fairbanks-Morse Co., Montreal.
A. B. Jardine & Co., Hespeler, Ont.
Trimont Mfg. Co., Roxbury, Mass.

Cutting Compound & Cutting Oil.
Can.l Economic Lubricant Co., Montreal.
Can. Oil Companies, Toronto.
Cataract Refining Co., Buffalo, N.Y.
Crescent Oil Co., New York.
Racine Tool & Machine Co., Racine, Wis.

Cutter Grinders and Attachments
Cincinnati Milling Machine Co., Cincinnati.
Garvin Machine Co., New York.
Girard Machine & Tool Co., Philadelphia, Pa.

Cutters, Milling.
A. R. Williams Machy. Co., Toronto
Can. Fairbanks-Morse Co., Montreal
Cleveland Twist Drill Co., Cleveland.
Garvin Machine Co., New York.
Morse Twist Drill and Machine Co., New Bedford.
H. W. Pettie, Toronto.
Tabor Mfg. Co., Philadelphia, Pa.
Pratt & Whitney Co., Dundas, Ont.
Wilt Twist Drill Co. of Canada, Ltd., Walkerville, Ont.

Cutting-off Machines.
Armstrong Bros. Tool Co., Chicago.
John Bertram & Sons Co., Dundas.
Can. Fairbanks-Morse Co., Montreal.
Espen-Lucas Machine Wks., Philadelphia, Pa.
Foss & Hill Machy. Co., Montreal.
Garlock-Machinery, Toronto.
Garvin Machine Co., New York.
Girard Machine & Tool Co., Philadelphia, Pa.
Geo. Gorton Machine Co., Racine, Wis.
Hurlbut, Rogers Machinery Co., South Sudbury, Mass.
John H. Hall & Sons, Brantford, Ont.
Wm. Kennedy & Sons, Owen Sound, Ont.
Nutter & Barnes Co., Hinsdale, N.H.
H. W. Petrie, Toronto.
Pratt & Whitney Co., Dundas, Ont
Tabor Mfg. Co., Philadelphia, Pa.
L. S. Starrett Co., Athol, Mass.

Damper Regulators.
Can. Fairbanks-Morse Co., Montreal.

Derricks.
Dominion Bridge Co., Montreal.
Wilt Twist Drill Co. of Canada, Ltd., Walkerville, Ont.

Designers, Special Machinery.
Baird Machine Co., Bridgeport, Conn.

Diamonds, Black.
Geo. A. Joyce Co., Ltd., New York.

Diamonds, Rough.
Geo. A. Joyce Co., Ltd., New York.

Diamond Tools.
Geo. A. Joyce Co., Ltd., New York.

Dies and Die Stocks.
Armstrong Mfg. Co., Bridgeport, Conn.
Banfield, W. H. & Son, Toronto.
Butterfield & Co., Rock Island, Que.
Brown, Boggs & Co., Hamilton, Ont.
Can. Fairbanks-Morse Co., Montreal.
Duncan Electrical Co., Montreal.
Gardner, Robt., & Son, Montreal.
Greenfield Tap & Die Corporation, Greenfield, Mass.
A. B. Jardine & Co., Hespeler, Ont.
Matthews, J. H., & Co., Pittsburg, Pa.
Modern Tool Co., Erie, Pa.
Morse Twist Drill and Machine Co., New Bedford.
H. W. Petrie, Toronto.
Pratt & Whitney Co., Dundas, Ont.
Wiley & Russell, Greenfield, Mass.

Dies for Bit Brace Use.
Wells Brothers Co., Greenfield, Mass.

Die Sinkers.
Garvin Machine Co., New York.

Dies for Machines.
Wells Brothers Co., Greenfield, Mass.

Die Sinking Presses, Hydraulic
Charles F. Elmes Eng. Works, Chicago
Watson-Stillman Co., Aldene, N.J.

Dies, Self-opening.
Geometric Tool Co., New Haven.
Greenfield Tap & Die Corporation, Greenfield, Mass.
Landis Machine Co., Waynesboro, Pa.
Matthews, J. H., & Co., Pittsburg, Pa.
Modern Tool Co., Erie, Pa.
Murchey Machine & Tool Co., Detroit

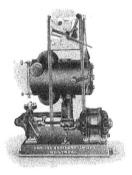
If what you want is not advertised in this issue consult the Buyers' Directory at the back.

Dies, Opening.
W. H. Banfield & Sons, Toronto.
Can. Fairbanks-Morse Co., Montreal.
Duncan Electrical Co., Montreal.
Greenfield Tap & Die Corporation, Greenfield, Mass.
A. B. Jardine & Co., Hespeler, Ont.
Landis Machine Co., Waynesboro, Pa.
Matthews, J. H., & Co., Pittsburg, Pa.
Modern Tool Co., Erie, Pa.
Murchey Machine & Tool Co., Detroit.
Pratt & Whitney Co., Dundas, Ont.

Dies for Screw Plates.
Wells Brothers Co., Greenfield, Mass.

Dies, Sheet Metal Working.
E. W. Bliss Co., Brooklyn, N.Y.
Brown, Boggs & Co., Hamilton, Can.

Dies, Screws and Thread.
Armstrong Mfg. Co., Bridgeport, Conn.
Greenfield Tap & Die Corporation, Greenfield, Mass.
Landis Machine Co., Waynesboro, Pa.
Modern Tool Co., Erie, Pa.
Murchey Machine & Tool Co., Detroit.

Discs, Leather.
Graton & Knight Mfg. Co., Montreal.

Draughtsman's Tools.
Emmert Mfg. Co., Waynesboro, Pa.

Drafts, Mechanical.
W. H. Banfield & Sons, Toronto.
Butterfield & Co., Rock Island, Que.
Can. Buffalo Forge Co., Montreal.
Can. Sirocco Co., Windsor, Ont.
A. B. Jardine & Co., Hespeler, Ont.
Pratt & Whitney Co., Dundas, Ont.
Sheldons, Limited, Galt, Ont.

Drill Bolt Cutters.
Cleveland Pneumatic Tool Co. of Canada, Toronto.

Drill Presses.
Baker Bros., Toledo, O.
W. F. & John Barnes Co., Rockford.
Can. Buffalo Forge Co., Montreal.
Colburn Machine Tool Co., Franklin, Pa.
Foss & Hill Machy. Co., Montreal.
Hill, Clarke & Co. of Chicago, Chicago, Ill.
Garvin Machine Co., New York.
Girard Machine & Tool Co., Philadelphia, Pa.
Niles-Bement-Pond Co., New York.
H. W. Petrie, Toronto.
A. R. Williams Machinery Co., Toronto.

Drilling Machines, Locomotive and Multiple Spindle.
Amalgamated Machy. Corporation, Chicago, Ill.
Baker Bros., Toledo, O.
Barnes Drill Co., Rockford, Ill.
John Bertram & Sons Co., Dundas.
Can. Buffalo Forge Co., Montreal.
Can. Fairbanks-Morse Co., Montreal.
Colburn Machine Tool Co., Franklin, Pa.
Garlock-Machinery, Toronto.
Garvin Machine Co., New York.
Girard Machine & Tool Co., Philadelphia, Pa.
A. B. Jardine & Co., Hespeler, Ont.
Niles-Bement-Pond Co., New York.
H. W. Petrie, Toronto.

Drilling Machines, Radial and Turret.
Baker Bros., Toledo, O.
Barnes Drill Co., Rockford, Ill.
John Bertram & Sons Co., Dundas.
Can. Fairbanks-Morse Co., Montreal.
Motch & Merryweather Machy. Co., Cleveland, O.
Niles-Bement-Pond Co., New York.
H. W. Petrie, Toronto.
Turner Machine Co., Danbury, Conn.

Drilling Machines, Sensitive.
Baker Bros., Toledo, O.
W. F. & John Barnes Co., Rockford.
Can. Fairbanks-Morse Co., Montreal.
Niles-Bement-Pond Co., New York.
Rockford Machine Tool Co., Rockford.

Drilling Machines, Upright and Horizontal.
Amalgamated Machy. Corporation, Chicago, Ill.
Baker Bros., Toledo, O.
Barnes Drill Co., Rockford, Ill.
Colburn Mach. Tool Co., Franklin, Pa.
A. R. Williams Machy. Co., Toronto.
W. F. & John Barnes Co., Rockford.
John Bertram & Sons Co., Dundas.
Garlock-Machinery, Toronto.
Girard Machine & Tool Co., Philadelphia, Pa.
A. B. Jardine & Co., Hespeler, Ont.
Rockford Machine Tool Co., Rockford.
R. McDougall Co., Galt.
Motch & Merryweather Machy. Co., Cleveland, O.
Niles-Bement-Pond Co., New York.
H. W. Petrie, Toronto.

Drilling Posts.
Keystone Mfg. Co., Buffalo, N.Y.

Drills, Bench.
W. F. & John Barnes Co., Rockford.
Can. Buffalo Forge Co., Montreal.
Can. Fairbanks-Morse Co., Montreal.
Pratt & Whitney Co., Dundas, Ont.
United States Electrical Tool Co., Cincinnati.

Drills, Blacksmith and Bit Stock.
Can. Buffalo Forge Co., Montreal.
Cleveland Twist Drill Co., Cleveland.
A. B. Jardine & Co., Hespeler, Ont.
Morse Twist Drill and Machine Co., New Bedford.
H. W. Petrie, Toronto.
Wilt Twist Drill Co., of Canada, Ltd., Walkerville, Ont.

Drills, Centre.
Cleveland Twist Drill Co., Cleveland.
Morse Twist Drill and Machine Co. New Bedford.
Pratt & Whitney Co., Dundas, Ont.
L. S. Starrett Co., Athol, Mass.
Wilt Twist Drill Co., of Canada, Ltd., Walkerville, Ont.

Drills Corner (Pneumatic).
Cleveland Pneumatic Tool Co. of Canada, Toronto.

Drills, Electric and Portable.
A. R. Williams Machy. Co., Toronto.
Can. Buffalo Forge Co., Montreal.
Niles-Bement-Pond Co., New York.
H. W. Petrie, Toronto.
Stow Mfg. Co., Binghamton, N.Y.
United States Electrical Tool Co., Cincinnati, O.

Drills, High Speed.
Baker Bros., Toledo, O.
Cleveland Twist Drill Co., Cleveland.
Can. Fairbanks-Morse Co., Montreal.
H. A. Drury Co., Montreal.
Morse Twist Drill and Machine Co., New Bedford.
W. F. & John Barnes Co., Rockford, Ill.
McKenna Bros. Brass Co., Pittsburg, Pa.
H. W. Petrie, Toronto.
Pratt & Whitney Co., Dundas, Ont.
Whitman & Barnes Mfg. Co., St. Catharines, Ont.
Wilt Twist Drill Co., of Canada, Ltd., Walkerville, Ont.

Drills, Multiple Spindle.
Pratt & Whitney Co., Dundas, Ont.
Niles-Bement-Pond Co., New York.

Drills, Oil Tube.
Cleveland Twist Drill Co., Cleveland.
Morse Twist Drill and Machine Co., New Bedford.

Drills, Pneumatic.
John F. Allen Co., New York.
Cleveland Pneumatic Tool Co. of Canada, Toronto.
Independent Pneumatic Tool Co., Chicago, Ill.
Niles-Bement-Pond Co., New York.

Drills, Ratchet and Hand.
Armstrong Bros. Tool Co., Chicago.
Can. Buffalo Forge Co., Montreal.
Can. Fairbanks-Morse Co., Montreal.
Cleveland Twist Drill Co., Cleveland.
A. B. Jardine & Co., Hespeler, Ont.
Morse Twist Drill and Machine Co., New Bedford.
H. W. Petrie, Toronto.
Pratt & Whitney Co., Dundas, Ont.
Wilt Twist Drill Co. of Canada, Ltd., Walkerville, Ont.

Drills, Rock.
A. R. Williams Machy. Co., Toronto.
Cleveland Pneumatic Tool Co. of Canada, Toronto.

Drills, Reamer.
McKenna Bros. Brass Co., Pittsburg, Pa.

Drills, Track.
Cleveland Twist Drill Co., Cleveland.
Morse Twist Drill and Machine Co., New Bedford.
Wilt Twist Drill Co. of Canada, Ltd., Walkerville, Ont.

Drills, Twist.
Armstrong, Whitworth of Canada, Ltd., Montreal.
Can. Fairbanks-Morse Co., Montreal.
Cleveland Twist Drill Co., Cleveland.
John Morrow Screw Co., Ingersoll, Ont.
Morse Twist Drill and Machine Co., New Bedford.
H. W. Petrie, Toronto.
Pratt & Whitney Co., Dundas, Ont.
Wilt Twist Drill Co. of Canada, Ltd., Walkerville, Ont.

Drill Holders.
Wells Brothers Co., Greenfield, Mass.

Drill Sockets.
Modern Tool Co., Erie, Pa.
Morse Twist Drill and Machine Co., New Bedford.
Wilt Twist Drill Co. of Canada, Ltd., Walkerville, Ont.

Drinking Fountains.
Puro Sanitary Drinking Fountain Co., Haydenville, Mass.

Drying Appliances.
Can. Buffalo Forge Co., Montreal.
Can. Sirocco Co., Ltd., Windsor, Ont.
Sheldons, Ltd., Galt, Ont.

Drying Out Barrels.
Baird Machine Co., Bridgeport, Conn.

Drying Ovens.
Oven Equipment & Mfg. Co., New Haven, Conn.
Whiting Foundry Equipment Co., Harvey, Ill.

Dump Cars.
Can. Fairbanks-Morse Co., Montreal.
National Machinery & Supply Co., Hamilton, Ont.
Owen Sound Iron Works Co., Owen Sound.
Plessisville Foundry, Plessisville, Que.

Dust Separators.
Can. Buffalo Forge Co., Montreal.
Sheldons Ltd., Galt, Ont.

Dust Arresters (for Tumbling Mills).
Northern Crane Works, Ltd., Walkerville, Ont.
Whiting Foundry Equipment Co., Harvey, Ill.

Dynamos and Electrical Supplies.
A. R. Williams Machy. Co., Toronto.
Can. Fairbanks-Morse Co., Montreal.
Lancashire Dynamo and Motor Co., Ltd., Toronto.
H. W. Petrie, Toronto.
Toronto & Hamilton Electric Co., Hamilton, Ont.

Electrical Supplies.
Duncan Electrical Co., Montreal.
Lintz-Porter Co., Toronto.

Elevator Enclosures.
Canada Wire & Iron Goods Co., Hamilton, Ont.
Dennis Wire & Iron Works, London, Ont.

Elevating and Conveying Machinery.
Can. Mathews Gravity Co., Toronto.
Plessisville Foundry, Plessisville, Que.

Emery Grinders (Pneumatic).
Cleveland Pneumatic Tool Co. of Canada, Toronto.
Stow Mfg. Co., Binghamton, N.Y.

Emery and Emery Wheels.
Can. Fairbanks-Morse Co., Montreal.
Canadian Hart Wheels, Hamilton, Ont.
Ford-Smith Machine Co., Hamilton.
Garvin Machine Co., New York.
Girard Machine & Tool Co., Philadelphia, Pa.
H. W. Petrie, Toronto.
Stevens, F. B., Detroit, Mich.

Emery Wheels, Dressers and Stands.
Canadian Hart Wheels, Hamilton, Ont.
Gardner, Robt., & Son, Montreal.
General Supply Co. of Canada, Ltd., Ottawa.
National Machinery & Supply Co. Hamilton, Ont.
Norton Co., Worcester, Mass.
H. W. Petrie, Toronto.

Emery Wheel Safety Flanges.
Canadian Hart Wheels, Hamilton, Ont.

Enameling Ovens.
Oven Equipment & Mfg. Co., New Haven, Conn.

Engines, Corliss, Compound, Pumping.
Mesta Machine Co., Pittsburg, Pa.
Wm. Tod Company, Youngstown, O.

Engines, Gas and Gasoline.
Can. Fairbanks-Morse Co., Montreal.
Jones & Glassco, Montreal.
Mesta Machine Co., Pittsburg, Pa.
National Machinery & Supply Co., Hamilton.
H. W. Petrie, Toronto.
Wm. Tod Company, Youngstown, O.

Engines, Horizontal and Vertical.
Can. Buffalo Forge Co., Montreal.
Can. Sirocco Co., Ltd., Windsor, Ont.
Mesta Machine Co., Pittsburg, Pa.
H. W. Petrie, Toronto.
Sheldons, Ltd., Galt, Ont.
A. R. Williams Machy. Co., Toronto.
Wm. Tod Co., Youngstown, O.

Engines, High-Speed, Automatic.
Can. Buffalo Forge Co., Montreal.

Engines, Steam.
Can. Buffalo Forge Co., Montreal.
General Supply Co. of Canada, Ltd., Ottawa.
Mesta Machine Co., Pittsburg, Pa.
H. W. Petrie, Toronto.
Plessisville Foundry, Plessisville, Que.
Southwark Foundry & Machine Co., Philadelphia, Pa.
Wm. Tod Company, Youngstown, O.

Engines, Stationary and Marine.
Southwark Foundry & Machine Co., Philadelphia, Pa.

Engineering Books.
The MacLean Publishing Co., Ltd., Toronto.

Engraving Machines.
Geo. Gorton Machine Co., Racine, Wis.

Elevators and Buckets.
Eastern Machinery Co., New Haven, Conn.
Whiting Foundry Equipment Co., Harvey, Ill.

Equipment Shop.
Baird Machine Co., Bridgeport, Conn.
Garvin Machine Co., New York.
Wm. Tod Co., Youngstown, O.

Escutcheon Pins.
Parmenter & Bulloch Co., Gananoque.

Evaporators' Machinery.
Brown, Boggs & Co., Hamilton, Can.

Exhaust Heads and Hoods.
Can. Buffalo Forge Co., Montreal.
Can. Steel Products Co., Montreal.
Can. Fairbanks-Morse Co., Montreal.
Sheldons, Ltd., Galt, Ont.

Exhausters.
Can. Buffalo Forge Co., Montreal.
Can. Sirocco Co., Ltd., Windsor, Ont.
H. W. Petrie, Toronto.

Experimental Machinery.
Owen Sound Iron Works Co., Owen Sound.

Extractors, Ingot.
Mesta Machine Co., Pittsburg, Pa.

Fans.
Can. Buffalo Forge Co., Berlin, Ont.
Baird Machine Co., Bridgeport, Conn.
Can. Sirocco Co., Ltd., Windsor, Ont.
Lintz-Porter Co., Toronto.
Plessisville Foundry, Plessisville, Que.
Sheldons, Ltd., Galt, Ont.
The Smart-Turner Machine Co., Hamilton.

Faucets.
Puro Sanitary Drinking Fountain Co., Haydenville, Mass.

Feed Water Heaters.
Can. Fairbanks-Morse Co., Montreal.
The Smart-Turner Machine Co., Hamilton.

Fence, Iron Factory.
Canada Wire & Iron Goods Co., Hamilton, Ont.
Dennis Wire & Iron Works Co., Ltd., London, Canada.
Standard Tube & Fence Co., Woodstock, Ont.

Files.
Delta File Works, Philadelphia, Pa.
Nicholson File Co., Port Hope, Ont.

Fire Alarm Apparatus.
Lintz-Porter Co., Toronto.

Fire Brick.
Elk Fire Brick Co., Hamilton, Ont.

Fire Extinguishers.
Can. H. W. Johns-Manville Co., Limited, Toronto.
General Supply Co. of Canada, Ltd., Ottawa.

Fire Escapes.
Canada Wire & Iron Goods Co., Hamilton, Ont.
Dennis Wire & Iron Works, London, Ont.

Flash Lamps.
Lintz-Porter Co., Toronto.

Flexible Shafts.
Chicago Flexible Shaft Co., Chicago.
Stow Mfg. Co., Binghamton, N.Y.

Flumes.
Toronto Iron Works, Ltd., Toronto.

Foot Valves.
Smart-Turner Mach. Co., Hamilton.

Forges, Hand, etc.
Can. Buffalo Forge Co., Montreal.
Independent Pneumatic Tool Co., Chicago, Ill.
National Machinery & Supply Co., Hamilton.
Sheldons, Limited, Galt, Ont.

Forgings, Drop, Automobile and Locomotive.
Bliss, E. W., Co., Brooklyn, N.Y.
Canadian Billings & Spencer, Ltd., Welland.
Dominion Forge & Stamping Co., Walkerville, Ont.
Mesta Machine Co., Pittsburg, Pa.
Steel Co. of Canada, Hamilton, Ont.
J. H. Williams Co., Brooklyn, N.Y.

Forging Hammers, Belt-Driven.
Bliss, E. W., Co., Brooklyn, N.Y.
Plessisville Foundry, Plessisville, Que.

Forging Hammers, Steam.
Erie Foundry Co., Erie, Pa.

Forging Machinery.
John Bertram & Sons Co., Dundas.
Bliss, E. W., Co., Brooklyn, N.Y.
Brown, Boggs Co., Ltd., Hamilton, Canada.
National Machinery Co., Tiffin, Ohio.
H. W. Petrie, Toronto.
Plessisville Foundry, Plessisville, Que.
Steel Co. of Canada, Hamilton, Ont.
Wm. Tod Company, Youngstown, O.
Watson-Stillman Co., Aldene, N.J.
Williams, White & Co., Moline, Ill.

Foundry Equipment.
Northern Crane Works, Walkerville.
W. W. Sly Mfg. Co., Cleveland, O.
Whiting Foundry Equipment Co., Harvey, Ill.

Friction Leathers.
Graton & Knight Mfg. Co., Montreal.

Friction Clutch Pulleys, etc.
American Pulley Co., Philadelphia, Pa.
Baird Machine Co., Bridgeport, Conn.
Eastern Machinery Co., New Haven, Conn.
Owen Sound Iron Works Co. Owen Sound.
H. W. Petrie, Toronto.
Positive Clutch & Pulley Works, Toronto.

Furnace Engineers and Contractors.
Mechanical Engineering Co., Montreal.
Whiting Foundry Equipment Co., Harvey, Ill.

Furnaces, Blast.
Toronto Iron Works, Ltd., Toronto.

Furnaces, Oil, Coal, Gas and Electric.
Canadian Hoskins, Limited, Walkerville, Ont.
Chicago Flexible Shaft Co., Chicago, Ill.
Mechanical Engineering Co., 55 Cote St., Montreal, Que.
H. W. Petrie, Toronto.
Tate, Jones & Co., Pittsburgh, Pa.
Whiting Foundry Equipment Co., Harvey, Ill.

Furnaces, Steel Heating and Brass Melting.
Can. Hoskins, Ltd., Walkerville, Ont.
Chicago Flexible Shaft Co., Chicago, Ill.
Mechanical Engineering Co., 55 Cote St., Montreal, Que.
Northern Crane Works, Ltd., Walkerville, Ont.
Tate, Jones & Co., Pittsburgh, Pa.
Whiting Foundry Equipment Co., Harvey, Ill.

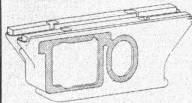

Furnaces, Heat Treating, Hardening and Tempering.
Can. Hoskins, Ltd., Walkerville, Ont.
Chicago Flexible Shaft Co., Chicago, Ill.
Mechanical Engineering Co., 55 Cote St., Montreal, Que.
Northern Crane Works, Ltd., Walkerville, Ont.
Tate, Jones & Co., Pittsburgh, Pa.
Whiting Foundry Equipment Co., Harvey, Ill.

Furnaces, Forging.
Mechanical Engineering Co., Montreal.
Northern Crane Works, Ltd., Walkerville, Ont.
Whiting Foundry Equipment Co., Harvey, Ill.

Furnaces, Annealing, etc.
Can. Hoskins, Ltd., Walkerville, Ont.
Chicago Flexible Shaft Co., Chicago, Ill.
Mechanical Engineering Co., 55 Cote St., Montreal, Que.
Northern Crane Works, Ltd., Walkerville, Ont.
Tate, Jones & Co., Pittsburgh, Pa.
Whiting Foundry Equipment Co., Harvey, Ill.

Furnaces for Baking, Bluing, Drying, Enameling, Japanning and Lacquering.
Mechanical Engineering Co., Montreal.
Oven Equipment & Mfg. Co., New Haven, Conn.

Furnace Lining.
Can. H. W. Johns-Manville Co., Limited, Toronto.
Mechanical Engineering Co., Montreal.

Fuse Boxes, Steel.
Dominion Forge & Stamping Co., Walkerville, Ont.

Fuse Cap Machinery.
Noble & Westbrook Mfg. Co., Hartford, Conn.

Gang Planer Tools.
Armstrong Bros. Tool Co., Chicago.

Gaskets, Leather, etc.
Graton & Knight Mfg. Co., Montreal.
Can. H. W. Johns-Manville Co., Limited, Toronto.

Gas Blowers and Exhausters.
Can. Buffalo Forge Co., Montreal.
Can. Sirocco Co., Ltd., Windsor, Ont.
Sheldons, Limited, Galt.
Southwark Foundry & Machine Co., Philadelphia, Pa.

Gas Burners.
Oven Equipment & Mfg. Co., New Haven, Conn.

Gas Machines.
Brown, Boggs & Co., Hamilton, Can.

Gas Producer Plants.
Can. Fairbanks-Morse Co., Montreal.

Gauges, Hydraulic Pressure.
Charles F. Elmes Eng. Works, Chicago
Watson-Stillman Co., Aldene, N.J.

Gauges, Standard.
Can. Fairbanks-Morse Co., Montreal.
Cleveland Twist Drill Co., Cleveland.
Greenfield Tap & Die Corporation, Greenfield, Mass.
Holden-Morgan Co., Toronto.
Morse Twist Drill and Machine Co., New Bedford.
Pratt & Whitney Co., Hartford, Conn.
Garvin Machine Co., New York.
National Machinery & Supply Co., Hamilton.
Southwark Foundry & Machine Co., Philadelphia.

Gear-cutting Machinery.
Girard Machine & Tool Co., Philadelphia, Pa.
Hamilton Gear & Machine Co., Toronto.
Hill, Clarke & Co., of Chicago, Chicago, Ill.
Motch & Merryweather Machy. Co., Cleveland, O.
National Machinery & Supply Co., Hamilton.
H. W. Petrie, Toronto.
Sheldons, Limited, Galt, Ont.
The Smart-Turner Machine Co., Hamilton.
Wm. Tod Co., Youngstown, O.
D. E. Whiton Machine Co., New London, Conn.
A. R. Williams Machy. Co., Toronto.

Gears, Cut, Mortise, Angle, Worm.
Gardner, Robt., & Son, Montreal.
Hamilton Gear & Machine Co., Toronto.
Hull Iron & Steel Foundries, Ltd., Hull, Quebec.
Jones & Glassco, Montreal, P.Q.
Wm. Kennedy & Sons, Ltd., Owen Sound, Ont.
Mesta Machine Co., Pittsburg, Pa.
Philadelphia Gear Works, Philadelphia, Pa.
Smart-Turner Machine Co., Hamilton, Ont.
Wm. Tod Co., Youngstown, O.

Gears, Rawhide.
Hamilton Gear & Machine Co., Toronto.
Gardner, Robt., & Son, Montreal.
Jones & Glassco, Montreal, P.Q.
Philadelphia Gear Works, Philadelphia, Pa.
Smart-Turner Machine Co., Hamilton, Ont.

Generators, Electric.
A. R. Williams Machy. Co., Toronto.
Can. Fairbanks-Morse Co., Montreal.
Lancashire Dynamo and Motor Co. Ltd., Toronto.
H. W. Petrie, Toronto.
Toronto and Hamilton Electric Co., Hamilton.

Grain for Polishing.
Norton Co., Worcester, Mass.

Graphite.
Can. H. W. Johns-Manville Co., Ltd., Toronto.
Jos. Dixon Crucible Co., Jersey City.
Stevens, F. B., Detroit, Mich.

Greases.
Can. Economic Lubricant Co., Montreal.

Grinders, Automatic Knife.
W. H. Banfield & Son, Toronto.

Grinders, Centre, Pedestal and Bench.
Canadian Hart Wheels, Ltd., Hamilton, Ont.
Cleveland Pneumatic Tool Co. of Canada, Toronto.
Ford-Smith Machine Co., Hamilton.
Foss & Hill Machy. Co., Montreal.
Gray Mfg. & Machine Co., Toronto.
Niles-Bement-Pond Co., New York.
Modern Tool Co., Erie, Pa.
Morse Twist Drill and Machine Co., New Bedford.
New Britain Machine Co., New Britain, Conn.
Norton Grinding Co., Worcester, Mass.
H. W. Petrie, Toronto.
Stow Mfg. Co., Binghamton, N.Y.
United States Electrical Tool Co., Cincinnati, O.

Grinders, Cutter.
Brown & Sharpe Mfg. Co., Providence, R.I.
Foss & Hill Machy. Co., Montreal.
Greenfield Machine Co., Greenfield, Mass.
H. W. Petrie, Toronto.
Pratt & Whitney Co., Dundas, Ont.

Grinders, Die Chaser.
Bignall & Keeler Mfg. Co., Edwardsville, Ill.
Landis Machine Co., Waynesboro, Pa.
Modern Tool Co., Erie, Pa.

Grinders, Disk.
Armstrong Bros. Tool Co., Chicago, Ill.
Gardner Machine Co., Beloit, Wis.
Norton Grinding Co., Worcester, Mass.

Grinders, Drill.
Garvin Machine Co., New York.
United States Electric Tool Co., Cincinnati, O.

Grinders, Cylinder, Internal.
Brown & Sharpe Mfg. Co., Providence, R.I.
Foss & Hill Machy. Co., Montreal.
Girard Machine & Tool Co., Philadelphia, Pa.
Grant Mfg. & Machine Co., Bridgeport, Conn.
Greenfield Machine Co., Greenfield, Mass.
Hill, Clarke & Co, of Chicago, Chicago, Ill.
Landis Tool Co., Waynesboro, Pa.
Modern Tool Co., Erie, Pa.
Motch & Merryweather Machy. Co., Cleveland, O.
Norton Grinding Co., Worcester, Mass.
H. W. Petrie, Toronto.

Grinders, Electric.
Lintz-Porter Co., Toronto.

Grinders, Pneumatic.
Cleveland Pneumatic Tool Co. of Canada, Toronto.
Independent Pneumatic Tool Co., Chicago, Ill.

Grinders, Portable, Electric, Hand, Tool, Post, Floor and Bench.
Baird Machine Co., Bridgeport, Conn.
Brown & Sharpe Mfg. Co., Providence, R.I.
Foss & Hill Machy. Co., Montreal.
Grant Mfg. & Machine Co., Bridgeport, Conn.
Greenfield Machine Co., Greenfield, Mass.
Hill, Clarke & Co, of Chicago, Chicago, Ill.
Landis Tool Co., Waynesboro, Pa.
Motch & Merryweather Machy. Co., Cleveland, O.
Norton Co., Worcester, Mass.
H. W. Petrie, Toronto.
United States Electrical Tool Co., Cincinnati.
A. R. Williams Machy. Co., Toronto.

Grinders, Swing, Portable, Electric.
United States Electrical Tool Co., Cincinnati.

Grinders, Tool and Holders.
Armstrong Bros. Tool Co., Chicago.
W. F. & John Barnes Co., Rockford, Ill.
Blount, J. G., & Co., Everett, Mass.
Brown & Sharpe Mfg. Co., Providence, R.I.
Greenfield Machine Co., Greenfield, Mass.
Hill Clarke & Co, of Chicago, Chicago, Ill.
Motch & Merryweather Machy. Co. Cleveland, O.
Tabor Mfg. Co., Philadelphia, Pa.

Grinders, Universal, Plain.
Girard Machine & Tool Co., Philadelphia, Pa.
Landis Tool Co., Waynesboro, Pa.
Modern Tool Co., Erie, Pa.

Grinders, Vertical Surface.
Brown & Sharpe Mfg. Co., Providence, R.I.
Can. Fairbanks-Morse Co., Montreal.
Girard Machine & Tool Co., Philadelphia, Pa.
Pratt & Whitney Co., Dundas, Ont.

Grinding and Polishing Machines, Portable, Pneumatic and Spring Frame.
Can. Fairbanks-Morse Co., Montreal.
Canadian Hart Wheels, Ltd., Hamilton, Ont.
Gardner, Robt., & Son, Montreal.
Garvin Machine Co., New York.
Girard Machine & Tool Co., Philadelphia, Pa.
Gray Mfg. & Machine Co., Toronto.
Greenfield Machine Co., Greenfield, Mass.
Hall & Sons, John H., Brantford.
Hill, Clarke & Co. of Chicago, Chicago, Ill.
Motch & Merryweather Machy. Co., Cleveland, O
Niles-Bement-Pond Co., New York.
Norton Co., Worcester, Mass.
H. W. Petrie, Toronto.
Stow Mfg. Co., Binghamton, N.Y.

Grinding Wheels.
Can. Fairbanks-Morse Co., Montreal.
Canadian Hart Wheels, Ltd., Hamilton, Ont.
Carborundum Co., Niagara Falls.
Ford-Smith Machine Co., Hamilton, Canada.
Gray Mfg. & Machine Co., Toronto.
Norton Co., Worcester, Mass.
H. W. Petrie, Toronto.

Guards, Window and Machine.
Canada Wire & Iron Goods Co., Hamilton, Ont.
Dennis Wire & Iron Works Co., Ltd., London, Canada.

Hack Saw Blades.
E. C. Atkins & Co., Hamilton, Ont.
Victor Saw Works, Ltd., Hamilton, Canada.
Diamond Saw & Stamping Works, Buffalo, N.Y.
Racine Tool & Machine Co., Racine, Wis.
L. S. Starrett Co., Athol, Mass.

Hack Saw Frames.
Ford-Smith Machine Co., Hamilton, Canada.
Garvin Machine Co., New York City.

Hammer High Speed.
High Speed Hammer Co., Rochester, N.Y.

Hammers, Drop and Belt Driven.
Bliss, E. W., Co., Brooklyn, N.Y.
Brown, Boggs Co., Ltd., Hamilton, Canada.
Canadian Billings & Spencer, Ltd., Welland.
A. B. Jardine & Co., Hespeler, Ont.
Girard Machine & Tool Co., Philadelphia, Pa.
National Machinery & Supply Co., Hamilton.
Niles-Bement-Pond Co., New York.
Plessisville Foundry, Plessisville, Que.
Toledo Machine & Tool Co., Toledo.

Hammers, Helve Power.
West Tire Setter Co., Rochester, N.Y.

Hammers, Pneumatic.
Cleveland Pneumatic Tool Co., of Canada, Toronto.

Hammers, Steam.
John Bertram & Sons Co., Dundas.
Girard Machine & Tool Co., Philadelphia, Pa.
National Machinery & Supply Co., Hamilton.
Niles-Bement-Pond Co., New York.

Hand Hoists & Trolleys.
Whiting Foundry Equipment Co., Harvey, Ill.

Hand Leathers or Pads.
Graton & Knight Mfg. Co., Montreal.

Hangers.
Baird Machine Co., Bridgeport, Conn.
Gardner, Robt., & Son, Montreal.
General Supply Co. of Canada, Ltd., Ottawa.
H. W. Petrie, Toronto.
The Smart-Turner Machine Co., Hamilton.

Hardness Testing Instruments.
Shore Instrument & Mfg. Co., New York.

Heating and Ventilating Engineers.
Can. Buffalo Forge Co., Montreal.
Can. Sirocco Co., Ltd., Windsor, Ont.
Sheldons, Ltd., Galt, Ont.

Heat Gauges, Hardening and Annealing.
Shore Instrument & Mfg. Co., New York.

Hide.
L. S. Tarshis & Sons, Montreal.

Hinge Machinery.
Baird Machine Co., Bridgeport, Conn.

Hinges.
London Bolt & Hinge Works, London, Ont.

Hoists, Hydraulic.
Southwark Foundry & Machine Co., Philadelphia.
Watson-Stillman Co., Aldene, N.J.

Hoisting and Conveying Machinery.
Beath, W. D., & Son, Toronto.
General Supply Co. of Canada, Ltd., Ottawa.
Northern Crane Works, Walkerville.
Owen Sound Iron Works Co., Owen Sound.
Southwark Foundry & Machine Co., Philadelphia.
Whiting Foundry Equipment Co., Harvey, Ill.

Hoists, Chain, Electric and Pneumatic.
Northern Crane Works, Walkerville.
Whiting Foundry Equipment Co., Harvey, Ill.

Hoists, Electric.
Northern Crane Works, Walkerville.
Whiting Foundry Equipment Co., Harvey, Ill.

Hoppers.
Toronto Iron Works, Ltd., Toronto.

Hose Clamp Tool.
Cleveland Pneumatic Tool Co. of Canada, Toronto.

Hose, Pneumatic.
Cleveland Pneumatic Tool Co., of Canada, Toronto.

Hose, Steam, Suction and Water.
Can. H. W. Johns-Manville Co., Limited, Toronto.
F. Reddaway & Co., Montreal.

Holders for Dies and Drills.
Wells Brothers, Company, Greenfield, Mass.
Wilt Twist Drill Co. of Canada, Ltd., Walkerville, Ont.

Horsehair.
L. S. Tarshis & Sons, Montreal.

Hydraulic Accumulators.
Can. Boomer & Boschert Press Co., Montreal.
Can. Fairbanks-Morse Co., Montreal.
Mesta Machine Co., Pittsburgh.
Niles-Bement-Pond Co., New York.
William R. Perrin, Ltd., Toronto.
The Smart-Turner Machine Co., Hamilton.
Southwark Foundry & Machine Co., Philadelphia.
Watson-Stillman Co., Aldene, N.J.

Hydraulic Machinery.
Can. Boomer & Boschert Press Co., Montreal.
Charles F. Elmes Eng. Works, Chicago.
Mesta Machine Co., Pittsburgh.
Niles-Bement-Pond Co., New York.
National Machinery & Supply Co., Hamilton.
William R. Perrin, Ltd., Toronto.
H. W. Petrie, Toronto.
Southwark Foundry & Machine Co., Philadelphia.
Wm. Tod Co., Youngstown, O.
Watson-Stillman Co., Aldene, N.J.
Wood, R. D., & Co., Philadelphia.

Indicators, Speed.
Brown & Sharpe Mfg. Co., Providence, R.I.
L. S. Starrett Co., Athol, Mass.

Index Centres.
Fred. C. Dickow, Chicago, Ill.
Garvin Machine Co., New York.

Ingot Metals.
A. C. Leslie & Co., Ltd., Montreal.

Intensifiers.
Mesta Machine Co., Pittsburg, Pa.
Southwark Foundry & Machine Co., Philadelphia.

Iron Filler.
Can. H. W. Johns-Manville Co., Ltd., Toronto.

Iron Ore.
Hanna & Co., M. A., Cleveland, O.

Jacks, Hydraulic.
Charles F. Elmes Eng. Works, Chicago
Southwark Foundry & Machine Co., Philadelphia.
Watson-Stillman Co., Aldene, N.J.

Jacks.
Can. Fairbanks-Morse Co., Montreal.
Northern Crane Works, Walkerville.
Norton, A. O., Coaticook, Que.
H. W. Petrie, Toronto.
Plessisville Foundry, Plessisville, Que.

Jacks, Pneumatic.
Northern Crane Works, Walkerville.

Jacks, Pit and Track.
Can. Fairbanks-Morse Co., Montreal.
Northern Crane Works, Walkerville.
Watson-Stillman Co., Aldene, N.J.

Japanning Ovens.
Oven Equipment & Mfg. Co., New Haven, Conn.

Jaws, Face Plate.
Cushman Chuck Co., Hartford, Conn.
Skinner Chuck Co., New Britain, Conn.

The advertiser would like to know where you saw his advertisement—tell him.

Jigs, Tools, etc.
- Hamilton Gear & Machine Co., Toronto.

Key Seaters.
- Baker Bros., Toledo, O.
- Garvin Machine Co., New York.
- Morton Mfg. Co., Muskegon Heights Mich.
- A. R. Williams Machy. Co., Toronto.

Kilns.
- Can. Buffalo Forge Co., Montreal.
- Sheldons, Limited. Galt, Ont.

Laboratories, Inspection and Testing.
- Can. Inspection & Testing Laboratories, Ltd., Montreal.

Lacquering Ovens.
- Oven Equipment & Mfg. Co., New Haven, Conn.

Ladles, Foundry.
- Northern Crane Works, Walkerville.
- Whiting Foundry Equipment Co., Harvey, Ill.

Lag Screw Gimlet Pointers.
- National Machy. Co., Tiffin, Ohio.

Lamps, Arc and Incandescent.
- Can. Fairbanks-Morse Co., Montreal.
- Can. H. W. Johns-Manville Co., Limited, Toronto.
- Ker & Goodwin. Brantford.

Lamps, Tungsten.
- Lintz-Porter Co., Toronto.

Lathe Chucks.
- Ker & Goodwin. Brantford.

Lathe Attachment for Shells.
- Lymburner, Ltd., Montreal.

Lathes, Automatic.
- Windsor Machine Co., Windsor, Vt.

Lathe Dogs and Attachments.
- Armstrong Bros. Tool Co., Chicago.
- Fay & Scott, Dexter, Maine.
- Hendey Machine Co., Torrington, Conn.
- National Forge & Tool Co., Erie, Pa.
- J. H. Williams Co., Brooklyn, N.Y.

Lathes, Bench.
- W. F. & John Barnes Co., Rockford.
- Blount, J. G., & Co., Everett, Mass.
- Can. Fairbanks-Morse Co., Montreal.
- Pratt & Whitney Co., Dundas, Ont.

Lathes, Band Turning.
- Jenckes Machine Co., Sherbrooke, Que.

Lathes, Engine.
- Amalgamated Machy. Corporation, Chicago, Ill.
- A. R. Williams Machy. Co., Toronto.
- W. F. & John Barnes Co., Rockford, Ill.
- John Bertram & Sons Co., Dundas.
- Can. Fairbanks-Morse Co., Montreal.
- Cincinnati Iron & Steel Co., Cincinnati, O.
- Fay & Scott, Dexter, Maine.
- Foss & Hill Machy. Co., Montreal.
- Gardner, Robt. & Son, Montreal.
- Garlock-Machinery, Toronto.
- Garvin Machine Co., New York.
- Girard Machine & Tool Co., Philadelphia, Pa.
- Hendey Machine Co., Torrington, Conn.
- Hill, Clarke & Co., of Chicago, Chicago, Ill.
- R. McDougall Co., Galt.
- Motch & Merryweather Machy. Co., Cleveland, O.
- Niles-Bement-Pond Co., New York.
- Oliver Machinery Co., Grand Rapids, Mich.
- H. W. Petrie, Toronto.
- Pratt & Whitney Co., Dundas, Ont.

Lathe Pans.
- New Britain Machine Co., New Britain, Conn.

Lathes, Patternmakers'.
- J. G. Blount Co., Everett, Mass.
- Fay & Scott, Dexter, Maine.
- Foss & Hill Machy. Co., Montreal.
- Garlock-Machinery, Toronto.
- H. W. Petrie, Toronto.

Lathes, Roll Turning.
- Mesta Machine Co., Pittsburgh.

Lathes, Screw Cutting.
- A. R. Williams Machy. Co., Toronto.
- John Bertram & Sons Co., Dundas.
- Cincinnati Iron & Steel Co., Cincinnati, O.
- Girard Machine & Tool Co., Philadelphia, Pa.
- Motch & Merryweather Machy. Co., Cleveland, O.
- Niles-Bement-Pond Co., New York.
- H. W. Petrie, Toronto.

Lathes, Spinning.
- Bliss E. W. Co., Brooklyn, N.Y.
- Toledo Mach. & Tool Co., Toledo, O.

Lathe, Turret and Speed.
- John Bertram & Sons Co., Dundas.
- Blount, J. G. & Co., Everett, Mass.
- Brown & Sharpe Mfg. Co., Providence, R.I.
- Can. Fairbanks-Morse Co., Montreal.
- Canada Machinery Corp., Galt, Ont.
- Cincinnati Iron & Steel Co., Cincinnati, O.
- Colburn Machine Tool Co, Franklin, Pa.
- Fay & Scott, Dexter, Maine.
- Foss & Hill Machy. Co., Montreal.
- Garlock-Machinery, Toronto.
- Garvin Machine Co., New York.
- Girard Machine & Tool Co., Philadelphia, Pa.
- Motch & Merryweather Machy. Co. Cleveland, O.
- New Britain Machine Co., New Britain, Conn.
- Niles-Bement-Pond Co., New York.
- Oliver Machinery Co., Grand Rapids, Mich.
- H. W. Petrie, Toronto.
- Pratt & Whitney Co., Dundas, Ont.
- Warner & Swasey Co., Cleveland, O.
- Windsor Machine Co., Windsor, Vt.
- A. R. Williams Machy. Co., Toronto.

Leather Strapping.
- Graton & Knight Mfg. Co., Montreal.

Lifts, Pneumatic.
- Whiting Foundry Equipment Co., Harvey, Ill.

Lighting Fixtures.
- Lintz-Porter Co., Toronto.

Link Belting.
- Can. Fairbanks-Morse Co., Montreal.
- Graton & Knight Mfg. Co., Montreal.
- Jones & Glassco, Montreal.

Linoleum Mill Machinery.
- Bertrams, Ltd., Edinburgh, Scotland.

Liquid Air.
- L'Air Liquide Society, Montreal, Toronto.
- Lever Bros., Toronto.

Lockers, Steel Wardrobe and Steel Material.
- Canada Wire & Iron Goods Co., Hamilton, Ont.
- Dennis Wire & Iron Works Co., Ltd., London, Canada.

Lockers.
- Canada Wire & Iron Goods Co., Hamilton, Ont.
- Dennis Wire & Iron Works Co., Ltd., London, Canada.

Locomotive Equipment.
- Can. Locomotive Co., Kingston, Ont.

Locomotives, Railroading, Contracting.
- Can. Locomotive Co., Kingston, Ont.
- National Machinery & Supply Co., Hamilton.

Lubricants.
- S. F. Bowser & Co., Fort Wayne, Ind.
- Can. Economic Lubricant Co., Montreal.
- Can. Oil Company, Toronto.
- Cataract Refining Co., Toronto.
- Can. Economic Lubricant Co., Montreal.

Machine Tools.
- Amalgamated Machy. Corporation, Chicago, Ill.
- Brown & Sharpe Mfg. Co., Providence, R.I.
- Can. Fairbanks-Morse Co., Montreal.
- Can. Machinery Corp., Galt, Ont.
- Garlock-Machinery, Toronto.
- General Supply Co. of Canada, Ltd., Ottawa.
- Modern Tool Co., Erie, Pa.
- Niles-Bement-Pond Co., New York.
- H. W. Petrie, Toronto.
- Pratt & Whitney Co., Dundas, Ont.
- J. H. Williams Co., Brooklyn, N.Y.

Machinery Dealers.
- Can. Fairbanks-Morse Co., Montreal.
- Garlock-Machinery, Toronto.
- Hill, Clarke & Co., of Chicago.
- Marshall & Huschart Machinery Co., Chicago.
- National Machinery & Supply Co., Hamilton.
- New York Machinery Exchange, New York.
- H. W. Petrie, Toronto.
- A. R. Williams Machy. Co., Toronto.

Machinery Guards.
- Jones & Glassco, Montreal, P.Q.
- Canada Wire & Iron Goods Co., Hamilton, Ont.
- A. R. Williams Machy. Co., Toronto.

Machinery Repairs.
- Cunningham & Sons, St. Catharines, Ont.
- Plessisville Foundry, Plessisville, Que.

Machinists' Scales, Small Tools and Supplies.
- Can. Fairbanks-Morse Co., Montreal.
- Frank H. Scott, Montreal.
- J. H. Williams & Co., Brooklyn, N.Y.

Magnetos.
- Lintz-Porter Co., Toronto.

Mandrels.
- Can. Fairbanks-Morse Co., Montreal.
- Cleveland Twist Drill Co., Cleveland.
- A. B. Jardine & Co., Hespeler, Ont.
- Morse Twist Drill and Machine Co., New Bedford.
- H. W. Petrie, Toronto.
- Pratt & Whitney Co., Dundas, Ont.
- Whitman & Barnes... Twist Drill Co. of Canada, Ltd., Walkerville, Ont.

Marine Engines.
- Cunningham & Sons, St. Catharines, Ont.

Marking Machinery.
- Brown, Boggs Co., Hamilton, Ont.
- Noble & Westbrook Mfg. Co., Hartford, Conn.

Marquises.
- Dennis Wire & Iron Works, London, Ont.

Measuring Tapes and Rules.
- James Chesterman & Co., Ltd., Sheffield, Eng.

Metallurgists.
- Can. Inspection & Testing Laboratories, Ltd., Montreal.
- Toronto Testing Laboratory, Ltd., Toronto.

Metals.
- L. S. Tarshis & Sons, Montreal.

Metal Cutting Machines.
- Hurlbut, Rogers Machinery Co., South Sudbury, Mass.
- Racine Tool & Machine Co., Racine.

Metal Stamping.
- Duncan Electrical Co., Montreal.

Meters, Electrical.
- Can. H. W. Johns Manville Co., Ltd., Toronto.
- Lintz-Porter Co., Toronto.

Mill Machinery.
- Cunningham & Sons, St Catharines.
- Alexander Fleck, Ltd., Ottawa.

Milling Attachments.
- John Bertram & Sons Co., Dundas.
- Brown & Sharpe Mfg. Co., Providence.
- Cincinnati Milling Machine Co., Cincinnati.
- Hendey Mach. Co., Torrington, Conn.
- Kempsmith Mfg. Co. Milwaukee, W.
- Mesta Machine Co., Pittsburg, Pa.
- Niles-Bement-Pond Co., New York.
- Pratt & Whitney Co., Dundas, Ont.
- Rockford Milling Machine Co., Rockford, Ill.

Milling Machines, Horizontal and Vertical.
- A. R. Williams Machy. Co., Toronto.
- Brown & Sharpe Mfg. Co., Providence.
- Hill, Clarke & Co. of Chicago, Chicago, Ill.
- John Bertram & Sons Co., Dundas.
- Foss & Hill Machy. Co., Montreal.
- Girard Machine & Tool Co., Philadelphia, Pa.
- Gooley & Edlund, Cortland, N.Y.
- Kempsmith Mfg. Co., Milwaukee, W.
- Motch & Merryweather Machy. Co., Cleveland, O.
- Niles-Bement-Pond Co., New York.
- H. W. Petrie, Toronto.
- Pratt & Whitney Co., Dundas, Ont.
- Rockford Milling Machine Co., Rockford, Ill.

Milling Machines, Plain, Bench and Universal.
- Brown & Sharpe Mfg. Co., Providence.
- Cincinnati Milling Machine Co., Cincinnati.
- Foss & Hill Machy. Co., Montreal.
- Garvin Machine Co., New York.
- Gooley & Edlund, Cortland, N.Y
- Hill, Clarke & Co., of Chicago, Chicago, Ill.
- Hendey Machine Co., Torrington.
- Kempsmith Mfg. Co., Milwaukee, Wis.
- Mesta Machine Co., Pittsburg, Pa.
- Motch & Merryweather Machy. Co., Cleveland, O.
- Niles-Bement-Pond Co., New York.
- H. W. Petrie, Toronto.
- Pratt & Whitney Co., Dundas, Ont.
- Rockford Milling Machine Co., Rockford, Ill.
- A. R. Williams Machy. Co., Toronto.

Milling Machines, Profile.
- Brown & Sharpe Mfg. Co., Providence.
- Can. Fairbanks-Morse Co., Montreal.
- Foss & Hill Machy. Co., Montreal.
- Garvin Machine Co., New York.
- Girard Machine & Tool Co., Philadelphia, Pa.
- Mesta Machine Co., Pittsburg, Pa.
- Motch & Merryweather Machy. Co., Cleveland, O.
- H. W. Petrie, Toronto.
- Pratt & Whitney Co., Dundas, Ont.

Milling Tools.
- Brown & Sharpe Mfg. Co., Providence.
- Geometric Tool Co., New Haven, Conn.
- Kempsmith Mfg. Co., Milwaukee, W.
- Tabor Mfg. Co., Philadelphia, Pa.

Mine Cars and Hitchings.
- Can. Fairbanks-Morse Co., Montreal.
- MacKinnon, Holmes Co., Sherbrooke.
- Modern Tool Co., Erie, Pa.
- Pratt & Whitney Co., Dundas, Ont.

Mining Machinery.
- A. R. Williams Machy. Co., Toronto.
- Can. Fairbanks-Morse Co., Montreal.
- Cleveland Pneumatic Tool Co., of Canada, Toronto.
- H. W. Petrie, Toronto.
- Toronto & Hamilton Electric Co., Hamilton, Ont.

Mixers, Hot Metal.
- Mesta Machine Co., Pittsburg, Pa.

Mortising Machines.
- Jones & Glassco, Montreal.

Motors, Electric.
- A. R. Williams Machy. Co., Toronto.
- Can. Fairbanks-Morse Co., Montreal.
- Lancashire Dynamo & Motor Co., Ltd., Toronto.
- Lintz-Porter Co., Toronto.
- Toronto & Hamilton Electric Co., Hamilton, Ont.

Motors, Pneumatic.
- Cleveland Pneumatic Tool Co. of Canada, Toronto.
- Independent Pneumatic Tool Co., Chicago.

Mufflers.
- Dominion Forge & Stamping Co., Walkerville, Ont.

Multiple Index Centres.
- Garvin Machine Co., New York.

Nipple Threading Machines.
- John H. Hall & Sons, Ltd., Brantford, Ont.
- Landis Machine Co., Waynesboro, Pa.

Nitrogen.
- L'Air Liquide Society, Montreal, Toronto.
- Lever Bros., Toronto.

Nozzles, Spray.
- Can. Buffalo Forge Co., Montreal.

Nuts, Semi-Finish and Finished.
- Galt Machine Screw Co., Galt, Ont.
- Steel Co. of Canada, Hamilton, Ont.

Nut Burring Machines.
- National Machy. Co., Tiffin, O.
- National Mach. & Sup. Co., Hamilton

Nut Machines (Hot).
- National Machy. Co., Tiffin, O.

Nut Facing and Bolt Shaving Machines.
- Garvin Machine Co., New York.
- National Machy. Co., Tiffin, O.
- National Mach. & Sup. Co., Hamilton

Nut Tappers.
- John Bertram & Sons Co., Dundas.
- Garvin Machine Co., New York.
- Greenfield Tap & Die Corporation, Greenfield, Mass.
- Hall, J. H., & Son, Brantford, Ont.
- A. B. Jardine & Co., Hespeler.
- Landis Machine Co., Waynesboro, Pa.
- National Machy. Co., Tiffin, O.
- National Mach. & Sup. Co., Hamilton

Nut Wrenches.
- Wells Brothers Co., Greenfield, Mass.

Oil Separators.
- Can. Fairbanks-Morse Co., Montreal.
- Sheldons, Ltd., Galt, Ont.
- Smart-Turner Machine Co., Hamilton.

Oil Stones.
- Carborundum Co., Niagara Falls, N.Y.
- Norton Co., Worcester Mass.

Ovens for Baking, Bluing, Drying, Enamelling, Japanning, and Lacquering.
- Geo. Gorton Machine Co., Racine, Wis.
- Oven Equipment & Mfg. Co., New Haven, Conn.
- Whiting Foundry Equipment Co. Harvey, Ill.

Oven Trucks, Steel.
- Oven Equipment & Mfg. Co., New Haven, Conn.

Ovens for Drying, Temper and Under Trucks.
- Oven Equipment & Mfg. Co., New Haven, Conn.

Overhead Systems.
- W. D. Beath & Son, Toronto.

Oscillating Valve Grinders (Pneumatic).
- Cleveland Pneumatic Tool Co. of Canada, Toronto.

Oxy-Acetylene Welding and Cutting Plants.
- L'Air Liquide Society, Montreal, Toronto.
- Lever Bros., Toronto.

Oxygen.
- L'Air Liquide Society, Montreal, Toronto.
- Lever Bros., Toronto.

Packings, Leather, Hydraulics, Etc.
- General Supply Co. of Canada, Ltd., Ottawa.
- Graton & Knight Mfg. Co., Montreal.
- William R. Perrin, Ltd., Toronto.
- H. W. Petrie, Toronto.
- Southwark Foundry & Machine Co., Philadelphia.

Packing, Rubber, etc.
- Can. H. W. Johns-Manville Co., Ltd., Toronto.

Pans, Lathe.
- Cleveland Wire Spring Co., Cleveland

Pans, Steel Shop.
- Cleveland Wire Spring Co., Cleveland

Paper Mill Machinery.
- Bertrams, Ltd., Edinburgh, Scotland.
- Can. Sirocco Co., Ltd., Windsor, Ont.

Partitions.
- Canada Wire & Iron Goods Co., Hamilton, Ont.
- Dennis Wire & Iron Works Co., Ltd., London, Canada.

Patent Solicitors.
- H. J. S. Dennison, Toronto.
- Fetherstonhaugh & Co., Ottawa.
- Marion & Marion, Montreal.
- Ridout & Maybee, Toronto.
- Ross Thomson & Co., Ottawa, Ont.
- Harold Shipman & Co., Ottawa.

Patterns.
- Galt Malleable Iron Co., Galt.
- Guelph Pattern Works, Guelph.
- Hamilton Pattern & Foundry Co., Hamilton, Ont.
- Owen Sound Iron Works Co., Owen Sound, Ont.
- Plessisville Foundry, Plessisville, Que.
- Toronto Pattern Works, Toronto.
- Wells Pattern & Machine Works, Toronto.

Patterns, Metal and Wood.
- Guelph Pattern Works, Guelph, Ont.

Pattern Shop Equipment.
- Oliver Machy. Co., Grand Rapids, Mich.

Perforated Metals and Ornamental Iron Goods.
- Canada Wire & Iron Goods Co., Hamilton.

Phosphor Bronze Castings.
- Tallman Brass & Metal Co., Hamilton.

Pickling Machines.
- Mesta Machine Co., Pittsburgh.

Pig Iron.
- Hanna & Co., M. A., Cleveland, O.
- Steel Co. of Canada, Hamilton, Ont.
- Stevens, F. B., Detroit, Mich.

Pinions, Mill Cut.
- Mesta Machine Co., Pittsburg, Pa.
- Wm. Tod Co., Youngstown, O.

EATH HOISTING ND CONVEYING MACHINERY

Overhead Runways and Trolleys,

Cranes,   Derricks,

Chain Blocks,

Electric Hoists and Trolleys,

Rope Blocks,

Friction Hoists,

Hydraulic and Hand Power Ash Hoists,

Coal Handling Machines,

Gravity Roller and Spiral Conveyors.

We Are Installing

BEATH OVERHEAD TRACKS, TROLLEYS AND HOISTS

For Hoisting and Conveying

5-in., 6-in., 8-in. and 9.2-in. Shells

in the receiving, forging, machinery and shipping departments. Beath Overhead Runways require no floor space and are particularly adapted for this service.

The weight of these Shells have caused a new problem in handling that will have to be met and overcome by manufacturers of these heavier types of explosives.

Let our engineering department show you how a Beath Overhead Runway can be made to fit into your requirements.

W. D. Beath & Son, Limited

ENGINEERS AND MANUFACTURERS

20 Cooper Avenue   -   TORONTO

EASTERN REPRESENTATIVES:
The A. M. Ellicott Co., 301 St. James St., Montreal

SPECIAL PRINGS

AND

crew Machine Products

MADE ON CONTRACT

the most exacting specifications, d deliveries made as desired.

Ask for Booklet 6-T.

Established 1857
THE WALLACE BARNES COMPANY
218 SOUTH STREET,   BRISTOL, CONN., U.S.A.
Man'frs of "Barnes-made" Products
Springs Screw Machine Products, Cold Rolled Steel and Wire

Making SHRAPNEL ?

Here is Standard Equipment

The Fay & Scott turret tool post shown here is being universally adopted as standard equipment for the manufacture of shrapnel.

The square head turret, style G, is used for turning the outside of the shell. We have made these turrets for years, and can fit them to any make or size of lathe, old or new.

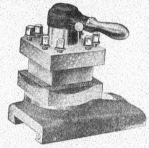

Style G
Catalog and full details on request

Fay & Scott, Dexter, Me.

If what you want is not advertised in this issue consult the Buyers Directory at the back.

Pipe Cutting and Threading Machines.
A. R. Williams Machy. Co., Toronto.
Armstrong Mfg. Co., Bridgeport, Conn.
Bignall & Keeler Mfg. Co., Edwardsville, Ill.
Butterfield & Co., Rock Island, Que.
Can. Fairbanks-Morse Co., Montreal.
Foss & Hill Machy. Co., Montreal.
Garvin Machine Co., New York.
Girard Machine & Tool Co., Philadelphia, Pa.
John H. Hall & Sons, Brantford.
A. B. Jardine & Co., Hespeler, Ont.
Landis Machine Co., Waynesboro, Pa.
R. McDougall Co., Galt.
H. W. Petrie, Toronto.
Trimont Mfg. Co., Roxbury, Mass.
Williams Tool Co., Erie, Pa.

Pipe Cutters, Rolling.
Armstrong Mfg. Co., Bridgeport, Conn.
Bignall & Keeler Mfg. Co., Edwardsville, Ill.
John H. Hall & Sons, Ltd., Brantford, Ont.

Pipe Fittings.
Southwark Foundry & Machine Co., Philadelphia.

Pipe, Riveted Steel.
Toronto Iron Works, Ltd., Toronto.

Pipe Straightening Machines.
Watson-Stillman Co., Aldene, N.J.

Planer Drives, Electrical.
Lancashire Dynamo & Motor Co., Ltd., Toronto.
Niles-Bement-Pond Co., New York.

Planer Jacks.
Armstrong Bros. Tool Co., Chicago.

Planers, Standard and Rotary.
John Bertram & Sons Co., Dundas.
Can. Fairbanks-Morse Co., Montreal.
Foss & Hill Machy. Co., Montreal.
Gardner, Robt., & Son, Montreal.
Garvin Machine Co., New York.
Girard Machine & Tool Co., Philadelphia, Pa.
Morton Mfg. Co., Muskegon Heights, Mich.
Niles-Bement-Pond Co., New York.
Oliver Machy. Co., Grand Rapids, Mich.
H. W. Petrie, Toronto.

Planing and Shaping Machinery.
A. R. Williams Machy. Co., Toronto.
Can. Fairbanks-Morse Co, Montreal.
Fay & Scott, Dexter, Maine.
Foss & Hill Machy. Co., Montreal.
Garvin Machine Co., New York.
Niles-Bement-Pond Co., New York.
H. W. Petrie, Toronto.

Planing Mill Exhausters.
Can. Buffalo Forge Co., Montreal.
Sheldons, Ltd., Galt, Ont.

Pliers.
Canadian Billings & Spencer, Ltd., Welland.

Pneumatic Tools.
Cleveland Pneumatic Tool Co. of Canada, Toronto.
Curtis Pneumatic Machinery Co., St. Louis, Mo.
Independent Pneumatic Tool Co., Chicago, New York.

Polishing Machines, Electric and Band.
Can. H. W. Johns-Manville Co., Toronto.

Portable Vise Stands.
New Britain Machine Co., New Britain, Conn.

Portable Steel Tool Racks.
New Britain Machine Co., New Britain, Conn.

Portable Steel Work Stands.
New Britain Machine Co., New Britain, Conn.

Power Plant Equipments.
Can. Fairbanks-Morse Co., Montreal.

Power Transmission.
Mesta Machine Co., Pittsburg, Pa.
The Smart-Turner Mach. Co., Hamilton.

Press Screw (Adjustable).
W. F. & John Barnes Co., Rockford.
Wm. R. Perrin, Ltd, Toronto.

Presses, Bench Straightening.
Toledo Machine & Tool Co., Toledo.

Presses for Shells.
Can. Boomer & Boschert Press Co, Montreal.
Can. Locomotive Co, Kingston, Ont.
Wm. Cramp & Sons Ship & Engine Building Co., Philadelphia, Pa.
Charles F. Elmes Eng. Works, Chicago
Foss & Hill Machy. Co., Montreal.
Goldie & McCulloch Co., Galt, Ont.
Mesta Machine Co., Pittsburgh.
William R. Perrin, Ltd, Toronto.
H. W. Petrie, Toronto.
Southwark Foundry & Machine Co, Philadelphia, Pa.
Wm. Tod Co., Youngstown, O.
Watson-Stillman Co., Aldene, N.J.
West Tire Setter Co, Rochester, N.Y.
Wood, R. D. & Co, Philadelphia.

Presses, Cam, Toggle, Eyelet.
Baird Machine Co., Bridgeport, Conn.
Toledo Machine & Tool Co., Toledo, O.

Presses, Broaching.
E. W. Bliss Co., Brooklyn, N.Y.
Toledo Machine & Tool Co. Toledo.
Watson-Stillman & Co., Aldene, N.J.

Presses, Drop.
W. H. Banfield & Son, Toronto.
E. W. Bliss Co, Brooklyn, N.Y.
Brown, Boggs Co., Ltd., Hamilton, Canada.
Can. Boomer & Boschert Press Co., Montreal.
Niles-Bement-Pond Co., New York.
William R. Perrin, Ltd., Toronto.
Toledo Machine & Tool Co., Toledo.
Watson-Stillman Co., Aldene, N.J.

Presses, Filter.
Lymburner, Ltd., Montreal.
Wm. R. Perrin, Ltd., Toronto.

Presses, Forging.
Can. Boomer & Boschert Press Co., Montreal.
E. W. Bliss Co., Brooklyn, N.Y.
Brown, Boggs Co., Ltd., Hamilton, Canada.
Wm. Cramp & Sons Ship & Engine Building Co., Philadelphia, Pa.
Charles F. Elmes Eng. Works, Chicago, Ill.
Can. Fairbanks-Morse Co., Montreal.
Girard Machine & Tool Co., Philadelphia, Pa.
Mesta Machine Co., Pittsburg, Pa.
Niles-Bement-Pond Co., New York.
Wm. R. Perrin, Ltd., Toronto.
H. W. Petrie, Toronto.
Southwark Foundry & Machine Co., Philadelphia, Pa.
Wm. Tod Co., Youngstown, O.
Toledo Machine & Tool Co., Toledo.
Watson-Stillman Co., Aldene, N.J.

Presses, Hydraulic.
Can. Boomer & Boschert Press Co., Montreal.
Wm. Cramp & Sons Ship & Engine Building Co., Philadelphia, Pa.
A. R. Williams Machy. Co., Toronto.
John Bertram & Sons Co., Dundas.
Charles F. Elmes Eng. Works, Chicago, Ill.
Mesta Machine Co., Pittsburg, Pa.
Niles-Bement-Pond Co., New York.
William R. Perrin, Ltd., Toronto.
Southwark Foundry & Machine Co., Philadelphia, Pa.
Wm. Tod Company, Youngstown, O.
H. W. Petrie, Toronto.
Toledo Machine & Tool Co., Toledo.
Watson-Stillman Co., Aldene, N.J.
Wood, R. D., & Co., Philadelphia.

Presses, Pneumatic.
Toledo Machine & Tool Co., Toledo.

Presses, Power.
Baird Machine Co., Bridgeport, Conn.
Can. Boomer & Boschert Press Co., Montreal.
E. W. Bliss Co., Brooklyn, N.Y.
Brown, Boggs & Co., Hamilton, Can.
Can. Fairbanks-Morse, Co., Montreal.
Charles F. Elmes Eng. Works, Chicago, Ill.
Geo. Gorton Machine Co., Racine.
Girard Machine & Tool Co., Philadelphia, Pa.
William R. Perrin, Ltd., Toronto.
H. W. Petrie, Toronto.
Southwark Foundry & Machine Co., Philadelphia, Pa.
Toledo Machine & Tool Co., Toledo.
Watson-Stillman Co., Aldene, N.J.
A. R. Williams Machy. Co., Toronto.

Presses, Scrap Baling.
Can. Boomer & Boschert Press Co., Montreal.
William R. Perrin, Ltd., Toronto.
Watson-Stillman Co., Aldene, N.J.

Presses, Spring Foot.
Baird Machine Co., Bridgeport, Conn.
Toledo Machine & Tool Co., Toledo.
Brown, Boggs & Co., Hamilton, Can.

Presses, Screw.
Can. Boomer & Boschert Press Co., Montreal.
Wm. R. Perrin, Ltd., Toronto.

Pressure Regulators.
Can. Fairbanks-Morse Co., Montreal.

Protective Paint.
Jos. Dixon Crucible Co., Jersey City.

Pulleys.
American Pulley Co., Philadelphia.
Baird Machine Co., Bridgeport, Conn.
Brown & Sharpe Mfg. Co., Providence, R.I.
Can. Fairbanks-Morse Co., Montreal.
General Supply Co. of Canada, Ltd., Ottawa.
Wm. Kennedy & Sons, Ltd., Owen Sound, Ont.
D. K. McLaren, Ltd., Montreal.
H. W. Petrie, Toronto.
Positive Clutch & Pulley Works, Ltd., Toronto.
The Smart-Turner Mach. Co., Hamilton.
A. R. Williams Machy. Co., Toronto.

Pulley Machinery, Drilling and Tapping.
Can. Fairbanks-Morse Co., Montreal.
Niles-Bement-Pond Co., New York.

Pumps, Air.
Mesta Machine Co., Pittsburg, Pa.
Smart-Turner Mach. Co., Hamilton.

Pumps, High Pressure.
Charles F. Elmes Eng. Works, Chicago
William R. Perrin, Ltd., Toronto.
Smart-Turner Mach. Co., Hamilton.
Southwark Foundry & Machine Co., Philadelphia.
Watson-Stillman Co., Aldene, N.J.

Pumping Machinery.
A. R. Williams Machy. Co., Toronto.
Can. Buffalo Forge Co., Montreal.
Can. Fairbanks-Morse Co., Montreal.
Darling Brothers, Montreal.
D'Olier Centrifugal Pump & Mach. Co., Philadelphia, Pa.
National Mach. & Sup. Co., Hamilton.
Wm. R. Perrin Co., Toronto.
H. W. Petrie, Toronto.
The Smart-Turner Mach. Co., Hamilton.
Southwark Foundry & Machine Co., Philadelphia.
Wm. Tod Company, Youngstown, O.

Pumps, all Kinds.
Can. Buffalo Forge Co., Montreal.
Charles F. Elmes Eng. Works, Chicago
Darling Brothers, Montreal.
General Supply Co. of Canada, Ltd., Ottawa.
Owen Sound Iron Works Co., Owen Sound.
William R. Perrin, Ltd., Toronto.
H. W. Petrie, Toronto.
The Smart-Turner Mach. Co., Hamilton.
A. R. Williams Machy. Co., Toronto.
Watson-Stillman Co., Aldene, N.J.

Pumps, Electrically Driven.
D'Olier Centrifugal Pump & Mach. Co., Philadelphia, Pa.
The Smart-Turner Mach. Co., Hamilton.

Pumps, Hydraulic.
Can. Boomer & Boschert Press Co., Montreal.
Charles F. Elmes Eng. Works, Chicago, Ill.
Darling Brothers, Montreal.
Smart-Turner Mach. Co., Hamilton.
Southwark Foundry & Machine Co., Philadelphia.
Wm. R. Perrin, Ltd., Toronto.
Wm. Tod Co., Youngstown, O.
Watson-Stillman Co., Aldene, N.J.

Pumps for Oiling Systems.
S. F. Bowser & Co., Fort Wayne, Ind.

Pumps, Steam.
Darling Brothers, Montreal.
Smart-Turner Mach. Co., Hamilton.
Wm. Tod Company, Youngstown, O.

Pump Leathers.
Graton & Knight Mfg. Co., Montreal.
Southwark Foundry & Machine Co., Philadelphia.

Punches and Dies.
W. H. Banfield & Sons, Toronto.
E. W. Bliss Co., Brooklyn, N.Y.
Brown, Boggs Co., Ltd., Hamilton, Canada.
Can. Buffalo Forge Co., Montreal.
Can. Fairbanks-Morse Co., Montreal.
Scott Bros., Halifax, Eng.
Gardner, Robt., & Son, Montreal.
Globe Machine & Stamping Co.
A. B. Jardine & Co., Hespeler, Ont.
H. W. Petrie, Toronto.
Pratt & Whitney Co., Dundas, Ont.
Toledo Machine & Tool Co., Toledo, O.

Punches, Power.
John Bertram & Sons Co., Dundas.
Bliss, E. W., Co., Brooklyn, N.Y.
Brown, Boggs Co., Ltd., Hamilton, Canada.
Girard Machine & Tool Co., Philadelphia, Pa.
Niles-Bement-Pond Co., New York.
Watson-Stillman Co., Aldene, N.J.

Punches, Pneumatic.
Jno. F. Allen Co., New York.

Punching Machines, Horizontal.
Bertrams, Ltd., Edinburgh, Scotland.
John Bertram & Sons Co., Dundas.
Bliss, E. W., Co., Brooklyn, N.Y.
Brown, Boggs Co., Ltd., Hamilton, Canada.
Long & Alstatter Co., Hamilton, Ohio.
Niles-Bement-Pond Co., New York.
Williams, White & Co., Moline, Ill.

Pyrometers.
Canadian Hoskins, Limited, Walkerville, Ont.
Shore Instrument & Mfg. Co., New York City.
Thwing Instrument Co., Philadelphia, Pa.

Quartering Machines.
John Bertram & Sons Co., Dundas.
Niles-Bement-Pond Co., New York.

Ratchet Wrenches.
Wells Brothers Co., Greenfield, Mass.

Railing, Iron and Brass.
Canada Wire & Iron Goods Co., Hamilton, Ont.
Dennis Wire & Iron Works Co., Ltd., London, Canada.

Rail Benders.
Niles-Bement-Pond Co., New York.

Railroad Tools.
Can. Fairbanks-Morse Co., Montreal.
Niles-Bement-Pond Co., New York.

Railroad Tools, Hydraulic.
Watson-Stillman Co., Aldene, N.J.

Rapping Plates.
Stevens, F. B., Detroit, Mich.

Ratchets.
Keystone Mfg. Co., Buffalo, N.Y.

Raw Hide Pinions.
Gardner, Robt., & Son, Montreal.
Hamilton Gear & Machine Co., Toronto.
Jones & Glassco, Montreal.
Smart-Turner Machine Co., Hamilton, Ont.

Reamers, Adjustable.
Can. Fairbanks-Morse Co., Montreal.
Cleveland Twist Drill Co., Cleveland
Morse Twist Drill & Machine Co., New Bedford.
Pratt & Whitney Co., Dundas, Ont.
Wells Brothers Co., Greenfield, Mass.

Reamers, Bridge, Expanding and High Speed.
Butterfield & Co., Rock Island, Que.
Can. Fairbanks-Morse Co., Montreal.
Cleveland Twist Drill Co., Cleveland.
McKenna Bros. Brass Co., Pittsburgh, Pa.
Morse Twist Drill & Machine Co., New Bedford.
H. W. Petrie, Toronto.
Pratt & Whitney Co., Dundas, Ont.

Reamer Fluting Machines.
Garvin Machine Co., New York.

Reamers, Pipe, Cylinder and Locomotive.
Butterfield & Co., Rock Island, Que.
Can. Fairbanks-Morse Co., Montreal.
Cleveland Twist Drill Co., Cleveland.
Morse Twist Drill & Machine Co., New Bedford.
Pratt & Whitney Co., Dundas, Ont.
Whitman & Barnes Mfg. Co., St. Catharines, Ont.
Wilt Twist Drill Co. of Canada, Ltd., Walkerville, Ont.

Reaming Machines, Pneumatic.
Cleveland Pneumatic Tool Co., of Canada, Toronto.
Independent Pneumatic Tool Co., Chicago.

Reamers, Steel Taper and Self-Feeding.
Butterfield & Co., Rock Island, Que.
Can. Fairbanks-Morse Co., Montreal.
Cleveland Twist Drill Co., Cleveland.
A. B. Jardine & Co., Hespeler, Ont.
Morse Twist Drill & Machine Co., New Bedford.
H. W. Petrie, Toronto.
Pratt & Whitney Co., Dundas, Ont.
Wilt Twist Drill Co., of Canada, Ltd., Walkerville, Ont.

Rebuilt Machine Tools.
New York Machy. Co., New York.

Reels.
Baird Machine Co., Bridgeport, Conn.

Rheostats.
Toronto & Hamilton Electric Co., Hamilton, Ont.

Rivet Machines.
Buffalo Forge Co., Buffalo, N.Y.
National Machinery Co., Tiffin, O.

Rivets, Tubular, Bifurcated.
Parmenter & Bulloch Co., Gananoque.

Rivets, Iron, Copper and Brass.
Parmenter & Bulloch Co., Gananoque.

Riveters, Pneumatic, Hydraulic.

Hammer, Compression.
Alliance Machine Co., Alliance, O.
Jno. F. Allen Co., New York.
Can. Fairbanks-Morse Co., Montreal.
Cleveland Pneumatic Tool Co. of Canada, Toronto.
Independent Pneumatic Tool Co. of Chicago, Ill.
Mesta Machine Co., Pittsburg, Pa.
National Mach. & Sup. Co., Hamilton
Niles-Bement-Pond Co., New York.
H. W. Petrie, Toronto.
Southwark Foundry & Machine Co., Philadelphia.
Watson-Stillman Co., Aldene, N.J.

Riveting Machines, Elastic Rotary Blow.
Girard Machine & Tool Co., Philadelphia, Pa.
Grant Mfg. & Machine Co., Bridgeport, Conn.
High-Speed Hammer Co., Rochester, N.Y.
F. B. Shuster Co., New Haven, Conn.
Southwark Foundry & Machine Co., Philadelphia.

Rolls, Bending.
John Bertram & Sons Co., Dundas, Ont.
Brown, Boggs Co., Ltd., Hamilton, Canada.
Niles-Bement-Pond Co., New York.
Toledo Machine & Tool Co., Toledo.

Rolling Mill Machinery.
Alliance Machine Co., Alliance, O.
Mesta Machine Co., Pittsburg, Pa.
Wm. Tod Co., Youngstown, O.

Roofing.
Can. H. W. Johns-Manville Co., Ltd., Toronto.

Rotary Converters.
A. R. Williams Machy. Co., Toronto.
Toronto and Hamilton Electric Co., Hamilton.

Rubbers.
L. S. Tarshis & Sons, Montreal.

Rubber Mill Machinery.
Bertrams, Ltd., Edinburgh, Scotland.
Can. H. W. Johns-Manville Co., Ltd., Toronto.

Rules.
Brown & Sharpe Mfg. Co., Providence, R.I.
James Chesterman & Co., Ltd., Sheffield, Eng.
L. S. Starrett Co., Athol, Mass.

Safety Set Screws.
Allen Mfg. Co., Inc., Hartford, Conn.

Sand Blasts.
Curtis Pneumatic Machinery Co., St. Louis, Mo.

THE "OLIVER" 16-INCH
HEAVY DUTY
ENGINE LATHE
POWERFUL
DOUBLE BACK GEARED
QUICK-CHANGE GEAR BOX
THREAD CUTTING
EARLY DELIVERIES
Write for Engine Lathe Bulletin No. 47
Write for Turret Lathe Bulletin No. 47T

Oliver Machinery Co.
Grand Rapids, Michigan, U.S.A.

Marking High Explosive Shells

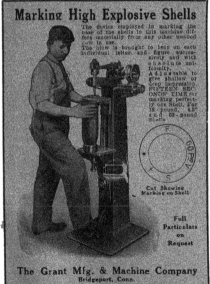

The device employed in marking the base of the shells in this machine differs materially from any other method now in use.

The blow is brought to bear on each individual letter and figure successively and with absolute uniformity.

Adjustable to give shallow or deep impression. FIFTEEN SECONDS' TIME for marking perfectly one Shell. For 18 - pound, 4.5 and 60 - pound Shells.

Cut Showing
Marking on Shell

Full
Particulars
on
Request

The Grant Mfg. & Machine Company
Bridgeport, Conn.

Motor Driven Speed Lathe

This style of motor drive employs a constant speed motor mounted on a plate having an extension arm to support a bearing for the outer end of the motor shaft. The motor plate is fitted to a slide on a shelf which is securely fastened to the back of the lathe bed. The motor plate is moved by means of a screw which tightens or loosens the belt. A four-step cone pulley on the motor shaft is belted to a four-step cone pulley on the spindle. This gives the same speed variation as when a countershaft is used, and by means of various size cones on the motor a wide range of speeds are obtainable.

The lathe spindle is made from high carbon steel, ground to size, and running in self-oiling bronze bearings. The tailstock has screw and lever feed. The bed is crossed-braced and all clamping levers are above the ways.

J. G. BLOUNT CO. - Everett, Mass., U.S.A.

Sand Blast Systems.
Whiting Foundry Equipment Co., Harvey, Ill.

Saw Blades.
Diamond Saw & Stamping Works, Buffalo, N.Y.

Sanding Machines.
Oliver Machy. Co., Grand Rapids, Mich.

Saw Tables.
Hub Machine Welding & Contracting Co., Philadelphia, Pa.

Saw Sharpening Machines.
Nutter & Barnes Co., Hinsdale, N.H.

Saw Mill Machinery.
A. R. Williams Machy. Co., Toronto
Can. Fairbanks-Morse Co., Montreal
Espen-Lucas Mach. Works, Philadelphia, Pa.
Gardner, Robt. & Son, Montreal.
Curtis Pneumatic Machinery Co., St. Louis, Mo.
National Mach. & Sup. Co., Hamilton.
H. W. Petrie, Toronto.
Plessisville Foundry, Plessisville, Que.

Saws, High-Speed, Friction.
Espen-Lucas Mach. Works, Philadelphia, Pa.
Hunter Saw & Machine Co., Pittsburg, Pa.
Mesta Machine Co., Pittsburg, Pa.
Nutter & Barnes Co., Hinsdale, N.H.

Saws, Inserted Tooth.
Espen-Lucas Mach. Works, Philadelphia, Pa.
Tabor Mfg. Co., Philadelphia, Pa.

Saws, Hack.
Can. Fairbanks-Morse Co., Montreal.
Diamond Saw & Stamping Works, Buffalo.
Ford-Smith Machine Co., Hamilton.
Garvin Machine Co., New York.
H. W. Petrie, Toronto.
L. S. Starrett Co., Athol, Mass.

Saws, Circular Metal.
H. A. Drury Co., Montreal.
Espen-Lucas Mach. Works, Philadelphia, Pa.
Hub Machine Welding & Contracting Co., Philadelphia, Pa.
Hunter Saw & Machine Co., Pittsburg, Pa.
Tabor Mfg. Co., Philadelphia, Pa.

Saws, Hot and Cold.
Hunter Saw & Machine Co., Pittsburg, Pa.
Mesta Machine Co., Pittsburgh.
Nutter & Barnes Co., Hinsdale, N.H.

Scleroscopes.
Shore Instrument & Mfg. Co., New York City.

Scrap Iron.
L. S. Tarshis & Sons, Montreal.

Screw Machine Products.
Wallace, Barnes Co., Bristol, Conn.

Screw Machines, Hand, Automatic.
Brown & Sharpe Mfg. Co., Providence, R.I.
Can. Fairbanks-Morse Co., Montreal.
Garvin Machine Co., New York.
Girard Machine & Tool Co., Philadelphia, Pa.
Hill, Clarke & Co., of Chicago, Chicago, Ill.
A. R. Jardine & Co., Hespeler.
Motch & Merryweather Machy. Co., Cleveland, O.
National Mach. & Sup. Co., Hamilton.
New Britain Machine Co., New Britain, Conn.
H. W. Petrie, Toronto.
Pratt & Whitney Co., Dundas, Ont.
Warner & Swasey Co., Cleveland, O.
A. R. Williams Machy. Co., Toronto.
Windsor Machine Co., Windsor, Vt.

Screw Machines, Multiple Spindle.
New Britain Machine Co., New Britain, Conn.
Windsor Machine Co., Windsor, Vt.

Screw Plates.
Butterfield & Co., Rock Island, Que.
Can. Tap & Die Co., Galt, Ont.
A. B. Jardine & Co., Hespeler.
Morse Twist Drill & Machine Co., New Bedford.
Wells Brothers Co., Greenfield, Mass.
Wiley & Russell Co., Greenfield, Mass.

Screw Slotters.
Garvin Machine Co., New York.
Pratt & Whitney Co., Dundas, Ont.

Set Screws, Safety.
Allen Mfg. Co., Hartford, Conn.

Second-Hand Machinery.
New York Machinery Co., New York.
Gardner, Robt. & Son, Montreal.
Can. Drawn Steel Co., Hamilton, Ont.
Gardner, Robt. & Son, Montreal.
National Mach. & Sup. Co., Hamilton.
Niles-Bement-Pond Co., New York.
H. W. Petrie, Toronto.
Plessisville Foundry, Plessisville, Que.
The Smart-Turner Machine Co., Hamilton.
Union Drawn Steel Co., Hamilton.

Shanks, Straight and Taper.
Jacobs Mfg. Co., Hartford, Conn.

Shapers.
John Bertram & Sons Co., Dundas.
Can. Fairbanks-Morse Co., Montreal.
Canada Machy. Corp., Galt, Ont.
Foss & Hill Machy. Co., Montreal.
Gardner, Robt., & Son, Montreal.
Girard Machine & Tool Co., Philadelphia, Pa.
Hendey Machine Co., Torrington, Ct.
Hill, Clarke & Co., of Chicago, Chicago, Ill.
H. W. Petrie, Toronto.

Shafting.
A. R. Williams Machy. Co., Toronto.
Can. Fairbanks-Morse Co., Montreal.
Mesta Machine Co., Pittsburg, Pa.
Niles-Bement-Pond Co., New York.
H. W. Petrie, Toronto.
Pratt & Whitney Co., Dundas, Ont.

Sharpening Stones.
Carborundum Co., Niagara Falls, N.Y.
Norton Co., Worcester, Mass.

Shavings, Separators.
Can. Buffalo Forge Co., Montreal.
Sheldons, Ltd., Galt, Ont.

Shearing Machines, Angle Iron, Bar and Gate.
John Bertram & Sons Co., Dundas.
Bertrams, Ltd., Edinburgh, Scotland.
Girard Machine & Tool Co., Philadelphia, Pa.
A. B. Jardine & Co., Hespeler.
Long & Alstatter, Hamilton, Ohio.
Mesta Machine Co., Pittsburg, Pa.
Niles-Bement-Pond Co., New York.
Scott Bros., Halifax, Eng.
Toledo Machine & Tool Co., Toledo.
Williams, White & Co., Moline, Ill.

Shears, Power.
John Bertram & Sons Co., Dundas.
Bliss, E. W., Co., Brooklyn, N.Y.
Brown, Boggs Co., Ltd., Hamilton, Canada.
Buffalo Forge Co., Buffalo, N.Y.
Girard Machine & Tool Co., Philadelphia, Pa.
Mesta Machine Co., Pittsburg, Pa.
National Machy. Co., Tiffin, Ohio.
National Mach. & Sup. Co., Hamilton.
Niles-Bement-Pond Co., New York.
Scott Bros., Halifax, Eng.
H. W. Petrie, Toronto.
Toledo Machine & Tool Co., Toledo.

Shears, Lever, Hydraulic.
Mesta Machine Co., Pittsburg, Pa.
Watson-Stillman Co., Aldene, N.J.

Shears, Pneumatic.
John F. Allen Co., New York.
Toledo Machine & Tool Co., Toledo, Ohio.

Shears, Squaring.
Brown, Boggs, & Co., Hamilton, Can.

Sheet Metal Working Tools.
Baird Machine Co., Bridgeport, Conn.
Bliss, E. W., Co., Brooklyn, N.Y.
Brown, Boggs & Co., Hamilton, Can.
Steel Bending Brake Works, Ltd., Chatham, Ont.

Sheet Metal Stampings.
Dominion Forge & Stamping Co., Walkerville, Ont.

Shell Banding Machines, Hydraulic.
Wm. Cramp & Sons Ship & Engine Bldg. Co., Philadelphia, Pa.
Can. Locomotive Co., Kingston, Ont.
Goldie & McCulloch Co., Galt, Ont.
Lymburner, Ltd., Montreal.
Motch & Merryweather Machy. Co., Cleveland, O.
Watson-Stillman Co., Aldene, N.J.
West Tire Setter Co., Rochester, N.Y.

Shell Hoisting Machinery.
Beath, W. D., & Son, Toronto.

Shell Lathes.
Barrett Machine Tool Co., Meadville, Pa.
Garlock-Machinery, Toronto.
Jenckes Machine Co., Sherbrooke, Que.
Kellogg & Co., Toronto.
H. W. Petrie, Toronto.

Shell Manufacturing Tools.
Amalgamated Machinery Corporation, Chicago, Ill.
Frank Toomey, Inc., Philadelphia, Pa.
Garlock-Machinery, Toronto.
New York Machinery Exchange, New York.
Hill, Clarke & Co. of Chicago.
H. W. Petrie, Toronto.

Shell Painting Machine.
Can. Buffalo Forge Co., Montreal.
Can. Locomotive Co., Kingston, Ont.

Shell Screws, Headless.
Blake & Johnson, Waterbury, Conn.

Shell Riveters.
Grant Mfg. & Machine Co., Bridgeport, Conn.

Shelving, Steel Partitions.
Canadian Steel Products Company, Montreal.

Sherardizing.
Chambers, Ltd., Toronto.

Shrapnel Shell Marker.
Brown-Boggs Co., Hamilton, Ont.
Holden-Morgan Co., Toronto.
Noble & Westbrook Mfg. Co., Hartford, Conn.

Shrapnel Sand Blasts.
W. W. Sly Mfg. Co., Cleveland, O.

Side Tools.
Armstrong Bros. Tool Co., Chicago.

Sirens, Electric.
Lintz-Porter Co., Toronto.
Sheldons, Ltd., Galt, Ont.

Silver Solder.
Geo. H. Lees & Co., Ltd., Hamilton, Ont.

Slotters.
Garvin Machine Co., New York.
Niles-Bement-Pond Co., New York.

Smokestacks.
MacKinnon, Holmes Co., Sherbrooke, Que.
Plessisville Foundry, Plessisville, Que.

Sockets.
Brown & Sharpe Mfg. Co., Providence, R.I.
Cleveland Twist Drill Co., Cleveland.
Keystone Mfg. Co., Buffalo, N.Y.
Modern Tool Co., Erie, Pa.
Morse Twist Drill & Machine Co., New Bedford.
Wilt Twist Drill Co. of Canada, Ltd., Walkerville, Ont.
Whitman & Barnes Mfg. Co., St. Catharines, Ont.
J. H. Williams Co., Brooklyn, N.Y.

Soldering Irons.
Brown, Boggs & Co., Hamilton, Can.

Solders.
Tallman Brass & Metal Co., Hamilton.

Specialties, Electric.
Lintz-Porter Co., Toronto.

Special Machinery.
Armstrong Bros., Toronto.
W. H. Banfield & Sons, Toronto.
John Bertram & Sons Co., Dundas.
Baird Machine Co., Bridgeport, Conn.
Bliss, E. W. Co., Brooklyn, N.Y.
Brown, Boggs & Co., Hamilton, Can.
Can. Fairbanks-Morse Co., Montreal.
Canada Machy. Agency, Montreal.
Cunningham & Sons, St. Catharines, Ont.
Charles F. Elmes Eng. Works, Chicago
Ford-Smith Machine Co., Hamilton.
Garvin Machine Co., New York.
Gooley & Edlund, Inc., Courtland, N.Y.
Grant Mfg. & Machy. Co., Bridgeport, Conn.
John H. Hall & Sons, Brantford.
Jardine, A. B., & Co., Hespeler.
National Electric Welder Co., Warren, Ohio.
National Forge & Tool Co., Erie, Pa.
National Mach. & Sup. Co., Hamilton.
Plessisville Foundry, Plessisville, Que.
Smart-Turner Machine Co., Hamilton, Ont.
William R. Perrin, Ltd., Toronto.
Wm. Tod Company, Youngstown, O

Spike Machines.
The Smart-Turner Machine Co., Hamilton.

Spring Coilers.
Baird Machine Co., Bridgeport, Conn.
Garvin Machine Co., New York.

Springs, Machinery.
Cleveland Wire Spring Co., Cleveland.
Jas. Steele, Ltd., Guelph, Ont.
Wallace, Barnes Co., Bristol, Conn.

Spring Making Machinery (Automatic).
Baird Machine Co., Bridgeport, Conn.

Sprockets, Chain.
Morse Chain Co., Ithaca, N.Y.
Philadelphia Gear Works, Philadelphia, Pa.

Stairs, Iron.
Canada Wire & Iron Goods Co., Hamilton, Ont.
Dennis Wire & Iron Works Co., Ltd., London, Canada.

Stampings.
Dominion Forge & Stamping Co., Walkerville, Ont.

Stamping Machinery.
Brown, Boggs & Co., Hamilton, Can.

Stationary Ladders.
New Britain Machine Co., New Britain, Conn.

Steam Specialties.
General Supply Co. of Canada, Ltd., Ottawa.
Sheldons, Ltd., Galt, Ont.

Steam Separators and Traps.
Can. Fairbanks-Morse Co., Montreal.
Can. Sirocco Co., Ltd., Windsor, Ont.
H. W. Petrie, Toronto.
Sheldons, Ltd., Galt, Ont.
The Smart-Turner Machine Co., Hamilton.

Steel Alloy.
Vanadium Alloys Steel Co., Pittsburgh, Pa.
Vulcan Crucible Steel Co., Aliquippa, Pa.

Steel Chains for Pulp Mill and Saw Mill.
Plessisville Foundry, Plessisville, Que.

Steel Barrels.
Smart-Turner Machine Co., Hamilton, Ont.

Steel Bench Legs.
New Britain Machine Co., New Britain, Conn.

Steel Bending Brakes.
Steel Bending Brake Works, Ltd., Chatham, Ont.

Steel, Cold Rolled.
Can. Drawn Steel Co., Hamilton, Ont.
A. C. Leslie & Co., Ltd., Montreal.
Union Drawn Steel Co., Hamilton, Ont.
Wallace, Barnes Co., Bristol, Conn.

Steel Drums.
Smart-Turner Machine Co., Hamilton, Ont.

Steel Pressure Blowers.
Can. Buffalo Forge Co., Montreal.
Can. Fairbanks-Morse Co., Montreal.

Steel, all kinds.
Lackawanna Steel Co., Lackawanna, N.Y.

Steel, High Speed.
Armstrong Whitworth of Canada, Ltd., Montreal.
Can. Fairbanks-Morse Co., Montreal.
H. A. Drury Co., Ltd., Montreal.
Thos. Firth & Sons, Montreal.
Hawkridge Bros. Co., Boston, Mass.
National Mach. & Sup. Co., Hamilton.
H. W. Petrie, Toronto.
Vanadium Alloys Steel Co., Pittsburg, Pa.
Vulcan Crucible Steel Co., Aliquippa, Pa.

Steel Die Engraving.
Noble & Westbrook Mfg. Co., Hartford, Conn.

Steel Machinery.
Hawkridge Bros. Co., Boston, Mass.

Steel Vanadium.
Vanadium Alloys Steel Co., Pittsburgh, Pa.
Vulcan Crucible Steel Co., Aliquippa, Pa.

Stock Racks for Bars, Piping, Etc.
New Britain Machine Co., New Britain, Conn.

Stocks for Dies.
Armstrong Manufacturing Co., Bridgeport, Conn.
Wells Bros. Co., Greenfield, Mass.

Stocks, Pipe.
Armstrong Manufacturing Co., Bridgeport, Conn.
Butterfield & Co., Rock Island, Que.
Greenfield Tap & Die Corporation, Greenfield, Mass.

Stools, Steel, Shop.
Dennis Wire & Iron Works Co., Ltd., London, Canada.

Storage Systems.
S. F. Bowser & Co., Fort Wayne, Ind.

Stoves, Electric.
Lintz-Porter Co., Toronto.

Straight Edges.
Steel Bending Brake Works, Ltd., Chatham, Ont.

Straightening Machinery.
Baird Machine Co., Bridgeport, Conn.
Bertrams, Ltd., Edinburgh, Scotland
National Mach. & Sup. Co., Hamilton.

Structural Steel.
Hamilton Bridge Works Co., Hamilton, Ont.
Lackawanna Steel Co., Lackawanna, N.Y.
Owen Sound Iron Works Co., Owen Sound, Ont.

Stud Driver.
Keystone Mfg. Co., Buffalo, N.Y.

Switchboards and Telephones.
Lintz-Porter Co., Toronto.
Toronto & Hamilton Electric Co., Hamilton.

Switches, Railway.
National Mach. & Sup. Co., Hamilton.

Tanks, Gasoline.
Dominion Forge & Stamping Co., Walkerville, Ont.

Tanks, Oil, Etc.
S. F. Bowser & Co., Fort Wayne, Ind.
MacKinnon, Holmes Co., Sherbrooke, Que.

Tanks, Steel.
John Inglis Co., Toronto.
MacKinnon, Holmes Co., Sherbrooke, Que.
Plessisville Foundry, Plessisville, Que.
Toronto Iron Works, Ltd., Toronto.

Tanks, Pressure.
Toronto Iron Works, Ltd., Toronto.

Tanks, Water.
MacKinnon, Holmes Co., Sherbrooke, Que.

Tank Wagons.
MacKinnon, Holmes Co., Sherbrooke, Que.
Toronto Iron Works, Ltd., Toronto.

Tapes, Measuring.
James Chesterman & Co., Ltd., Sheffield, Eng.

Tapes, Friction.
Can. H. W. Johns-Manville Co., Ltd., Toronto.

Tapping Machines (Pneumatic).
Cleveland Pneumatic Tool Co. of Canada, Toronto.
Independent Pneumatic Tool Co., Chicago, Ill.

Tapping Machines and Attachments.
Baker Brothers, Toledo, O.
John Bertram & Sons Co., Dundas.
Garvin Machine Co., New York.
The Geometric Tool Co., New Haven.
Girard Machine & Tool Co., Philadelphia, Pa.
Greenfield Tap & Die Corporation, Greenfield, Mass.
J. H. Hall & Sons, Brantford, Ont.
A. B. Jardine & Co., Hespeler.
Landis Machine Co., Waynesboro, Pa.
Manufacturers Equipment Co., Chicago, Ill.
Modern Tool Co., Erie, Pa.
Murchey Machine & Tool Co., Detroit.
Niles-Bement-Pond Co., New York.
H. W. Petrie, Toronto.
Rickart Shaper Co., Erie, Pa.
L. S. Starrett Co., Athol, Mass.

A Sensible Suggestion For You

With Christmas but three weeks away, our thoughts naturally turn to the time-honored custom of giving gifts of remembrance to our friends.

It has been a year of serious thinking, and the thoughts of the nation will be reflected in its Christmas giving. The useful gift will be the most acceptable and the most appreciated.

Let us suggest something that, considering its real value, will prove comparatively inexpensive.

Something that will constantly remind the recipient of your thoughtfulness.

Something that will prove a neat compliment to the one receiving it, that you considered him capable of appreciating a gift of this character.

Christmas Greetings

At the direction of

you have been entered upon our subscription list to receive

for one year.

It is our hope that each copy you receive may serve as a pleasant reminder of the one who sends you this holiday remembrance.

The MacLean Publishing Co., Limited, Toronto.

Let Us Suggest Canadian Machinery

Give **CANADIAN MACHINERY** to your employees and to your friends this Christmas.

It is only $2.00 for 52 issues, yet throughout the year its value will be magnified as its usefulness becomes more fully appreciated.

Send us the list of names and addresses, and we will send a handsome three-colored announcement card, a small reproduction of which is shown. This, together with the first copy of CANADIAN MACHINERY, will reach the recipient on Christmas Day. *Try it this year!*

CANADIAN MACHINERY

143 University Avenue, Toronto, Ontario, Canada

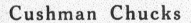
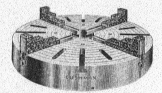

If what you want is not advertised in this issue consult the Buyers' Directory at the back.

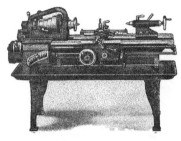
INDEX TO ADVERTISERS

The advertiser would like to know where you saw his advertisement—tell him.

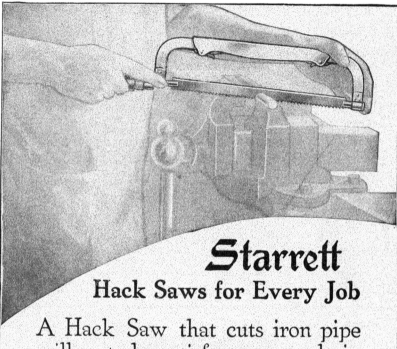

Starrett
Hack Saws for Every Job

A Hack Saw that cuts iron pipe will not do satisfactory work in cutting sheet steel.

Good machinists know this and also know that there is a Starrett blade for every purpose. Seventeen different blades— some differ in shape and size of teeth—some in composition, some in hardness. Our free catalog tells what saw to use. Guided by this you can recommend the proper blade for any use and give real service to your customers.

We deal direct with Hardware Stores
Send for free Catalog No. 20- 3 — prices and terms

The L. S. Starrett Company, Athol. Mass.
"The World's Greatest Tool Makers"

New York   London   Chicago

Lightning Source UK Ltd.
Milton Keynes UK
UKHW020751251118
332796UK00002B/32/P